The World of
Butterflies & Moths

The World of Butterflies & Moths

Umberto Parenti

ORBIS PUBLISHING · London

English language edition prepared in consultation with: Dr J. D. Bradley of the Commonwealth Institute of Entomology, London, and Allan Watson of the Natural History Museum, London.

Title page : Red Admiral, Vanessa atalanta *(Nymphalidae), a migratory butterfly resident in Europe and North America.*
Endpapers : A pair of Citrus Swallowtails, Papilio demodocus *(Papilionidae), which is frequently found round shrubs in central South Africa.*

Acknowledgements
Front cover: Bruce Coleman-Jane Burton; Endpapers: H. Jacana; I. Bucciarelli 2–3; S.E.F. 6; S.E.F. 8a; S.E.F. 8b; I.G.D.A. 9a; S.E.F. 9b; Bruce Coleman-Des & Jen Bartlett 10; Jacana-H. Chaumeton 12; Jacana-M. Moiton 13; Jacana-K. Ross 14; Jacana-M. Moiton 15; Jacana-K. Ross 16; Jacana-Pilloud 17a; Jacana-M. Moiton 17b; Bruce Colemann-Burton 18; Jacana-M. Moiton 19; Ricciorini-Giussani 20; Bruce Coleman-Burton 21; Bruce Coleman-Burton 22; Jacana-M. Moiton 23; Jacana-H. Chaumeton 24–25; Jacana H. Chaumeton 26a; Jacana-H. Chaumeton 26b; Jacana-Yolf 27; Jacana-M. Moiton 28; Jacana-M. Moiton 29a; Jacana-M. Chaumeton 29b; Jacana-A. Ducrot 31; Jacana-H. Chaumeton 32a; Jacana-H. Chaumeton 32b; Jacana-M. Moiton 34; D. Vaughan 35; Jacana-M. Moiton 36; Jacana-Gillon 38; Jacana-M. Moiton 39; Jacana-H. Chaumeton 40–41; Jacana-P. Dupont 41; I. Bucciarelli 42; Jacana-H. Chaumeton 43; Jacana-H. Chaumeton 44; Jacana-M. Moiton 45; I. Bucciarelli 46; Jacana-H. Chaumeton 47; Jacana-H. Chaumeton 48; Jacana-H. Chaumeton 49; Jacana-M. Moiton 50a; Jacana-M. Moiton 50b; Jacana-M. Moiton 51; I. Bucciarelli 52; Jacana-M. Moiton 53; Jacana-M. Moiton 54a; Jacana-M. Moiton 54b; Jacana-M. Moiton 55; Jacana-M. Moiton 56a; Jacana-M. Moiton 56b; Jacana-M. Moiton 56c; Jacana-M. Moiton 56–57; Jacana-H. Chaumeton 58a; I. Bucciarelli 58b; Jacana-M. Moiton 59; Jacana-M. Moiton 60; Jacana-H. Chaumeton 62; I. Bucciarelli 63; Jacana-G. Haüsle 64; Jacana-A. Visage 65; Jacana-K. Ross 66; Jacana-M. Moiton 67; I. Bucciarelli 68a; Jacana-M. Moiton 68b; Jacana-M. Moiton 69; Jacana-J. Labat 70–71; Jacana-Noailles 72; Jacana-Noailles 73; Jacana-Ducrot 74; Bruce Coleman 75; I. Bucciarelli 76a; I.G.D.A.-Bevilacqua 76b; Jacana-M. Moiton 77a; I. Bucciarelli 77b; Jacana-Bassot 78; Jacana-K. Ross 79; Jacana-H. Chaumeton 80; Ricciorini-Giussani 82; Jacana-K. Ross 83; Jacana-Nardin 84; I. Bucciarelli 85; Jacana-K. Ross 86; Jacana-M. Moiton 87; Ricciorini-Giussani 88; Ricciorini-Giussani 89; Ricciorini-Giussani 91; Jacana-Rouxaime 92; I. Bucciarelli 93a; Jacana-A. Ducrot 93b; Jacana-M. Moiton 94; Jacana-M. Moiton 95; I. Bucciarelli 96; I. Bucciarelli 97; I. Bucciarelli 98; Jacana-M. Moiton 99; I. Bucciarelli 100; Jacana-G. Haüsle 101; I. Bucciarelli 102; I. Bucciarelli 103; I. Bucciarelli 104; Jacana-H. Chaumeton 105; Jacana-M. Moiton 106; I. Bucciarelli 107; I. Bucciarelli 108a; I. Bucciarelli 108b; I. Bucciarelli 109; Bruce Coleman-B. Coates 110–111; Jacana-M. Moiton 112; Jacana-H. Chaumeton 113; Jacana-H. Chaumeton 114; I. Bucciarelli 115; I. Bucciarelli 116; Jacana-H. Chaumeton 117; Jacana-Ermie 118; D. Vaughan 119a; I. Bucciarelli 119b; I. Bucciarelli 119c; I. Bucciarelli 120; Jacana-H. Chaumeton 121; Jacana-C. de Klemm 122a; D. Vaughan 122b; Jacana-H. Chaumeton 123; Jacana-M. Moiton 124; Jacana-Rouxaime 125. The drawings on pages 13 and 43 are by Brian Hargreaves.

Contents

Introduction

Butterflies and moths are some of the most successful insects in colonizing the earth. They are found in all types of environment from Arctic tundra and arid deserts to lush Equatorial forests and from sea-level to the tops of mountains. Like man they do not actually live at the earth's poles or on the icy summits of the highest mountains, although some do inhabit remote, windswept oceanic islands. A few are aquatic or partially so, living in ponds and lakes, the backwaters of streams and rivers, or in coastal marshes and swamps, but so far as is known none is truly maritime. The majority are dependent on flowering plants, which are themselves dominant in the plant kingdom.

This book looks at some of the many facets of the world of butterflies and moths. One of the greatest attractions of these insects is their fantastic variety of shapes and colour-patterns, and something of this immense visual appeal is conveyed in the colour photographs which are reproduced, mostly showing the living insects larger than life size. Their evolution, classification, adaptation to different environments, four-stage life-cycle, anatomy and reproduction are among the aspects considered, as well as something of their interrelationship with man in their role of useful insects or as pests.

Butterflies and moths are classified scientifically as Lepidoptera. This name is derived from the Greek words *lepis*, meaning a scale, and *pteron*—a wing; it refers to the minute overlapping scales on the wings of the adult or imago. The presence of these scales and the coiled tongue or proboscis, which is developed in most Lepidoptera for taking liquid nourishment, distinguishes them from other insects. The vast majority of Lepidoptera are associated with flowering plants, and the particular plant or plants on which they feed as caterpillars are referred to as foodplants

or hostplants. Relatively few are non-vegetarian and live on animal matter.

Butterflies and moths differ from one another in several fundamental ways. Moths usually have feathery, tapering, or hair-like antennae, while butterfly antennae end in small knobs. Moths mainly alight with their wings stretched out flat, while butterflies generally rest with their wings held upright over their bodies. Butterflies have slender bodies with a few smooth, short scales; moths on the other hand often have stouter 'furry' bodies.

More than 160,000 species of Lepidoptera, about 15,000 of them butterflies, are known to entomologists. It is thought that there are certainly many more moths which await discovery. Even in such well-explored parts of the world as Europe, North America and Australia new forms are continually being found. Many rare species are known from only one or two specimens, while on the other hand little or nothing is known about the life histories of some of the common species. The world of butterflies and moths thus offers plenty of opportunities for original field study, and for those who may feel inspired to learn more about them a bibliography is appended.

Lepidoptera inhabited the earth long before man, and their origin and fossil history is intriguing in spite of being largely a matter of conjecture. Fossil specimens are rare and only recently have any been found to prove that Lepidoptera existed earlier than the Tertiary period about 40 million years ago. In 1970 a report was published describing the head capsule of a caterpillar found in Canadian amber of the Cretaceous period (about 70 million years ago) which had been collected from an open-pit coal mine near Medicine Hat, Alberta. Since then the fossil record has moved 20 or 30 million years further back following the discovery of some micro-moths preserved in Lebanese amber of the Lower Cretaceous period (more than 100 million years ago). These fossil moths are of very small size and it is perhaps significant that no monster dinosaurian Lepidoptera are known from the earlier Mesozoic period, (more than 150 million years ago), and that all the fossils so far discovered are smaller than the Birdwing butterflies

Left: Plate No. 196 (dated 1779) of Volume III of Pierre Cramer's rare work Papillons exotiques des trois parties du monde, l'Asie, l'Afrique et l'Amerique. *This plate shows the upper and lower side of three species of butterfly named by Cramer as follows: A and B—Papilio Erminia, found in Bengal; C and D—Papilio Leda, also found in Bengal and E and F—Papilio dido found in Surinam.*

and Atlas moths that live in the forests of Malaysia and northern Australia today. The close interrelationship of the Lepidoptera with flowering plants, which fossil evidence shows evolved in the Cretaceous period, suggests that this was also the time of Lepidoptera expansion.

Although evolution and the development of new forms—the process of speciation—never ceases, it seems likely that the Lepidoptera have already reached or passed their peak in number of species, since many are becoming increasingly scarce or extinct as their habitats are claimed by man for his own use. The pollution of the environment with insecticides, herbicides and the waste products of industry also takes its toll. Thankfully there is a growing awareness throughout the world of these dangers to animal and plant life, and of the need for conservation by creating national parks and reservations to give wildlife a better chance of survival.

The systematic classification (taxonomy) and naming (nomenclature) of all animals is founded on the work of the Swedish naturalist Carl von Linné, whose name is often latinized to Linnaeus. His book the *Natural System* was published in Latin, which was the most widely understood language of the mid-eighteenth century. Latin has since been adopted internationally for the names of animals and plants, and all names are latinized even if they are not actually founded on Latin words. The system of naming species is binominal; that is to say that each species has two basic names, the first being the generic name (genus) which always starts with a capital letter, and the second, the specific name (species), which starts with a small letter. These two names are generally printed in italics, and together they form the scientific name of a species, for example, *Papilio machaon*. They are usually accompanied by the name of the person, (not in italics), who first described the species, as in *Papilio machaon* Linnaeus. When a species is divided into different races or isolated populations a subspecific name may be added, so that the scientific name then consists of three parts: *Papilio machaon britannicus*. The higher categories, above generic levels, most frequently used in the Lepidoptera are, in ascending order for example for the papilionids: subfamily (Papilioninae), family (Papilionidae), superfamily (Papilionoidea), suborder (Ditrysia) and order (Lepidoptera).

Many Lepidoptera have common or colloquial names for *Papilio machaon* this is the Swallowtail butterfly. Such names are often very apt and attractive, but they can be very confusing as they tend to vary from place to place or have changed in the course of time. Many of the common names used today first appeared in the works of the English entomologists John Curtis and Moses Harris towards the end of the eighteenth century. Some of them were delightful appellations, euphonius and descriptive. Among those which have changed with time are The Lady of the Woods which today is known as the Orange-tip butterfly, *Anthocharis cardamines* (Pieridae), and The Bee Tyger moth which is now known as the Death's-head Hawk-moth, *Acherontia atropos* (Sphingidae); the Convolvulus Hawk-moth, *Agrius*

Below left: The pre-imaginal stages of the Giant Emperor Moth, Saturnia pyri (Saturniidae) of Europe, as illustrated in Duponchel's L'Iconographie des Chenilles.
Below right: Plate No 6 by Achilles Costa, from the 1857 work Degl'Insetti che attaccano l'albero e il frutto dell'olivo, del ciliegio, del pero, del castagno e della vite ecc. which shows the insects that attack the trees and fruits of the olive, the cherry, the pear, the chestnut and the vine.

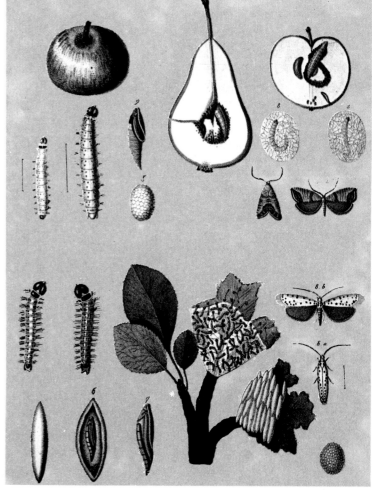

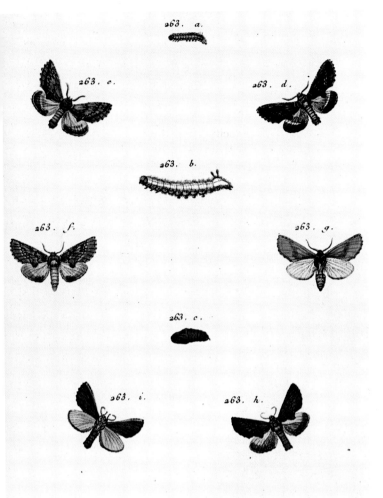

convolvuli, was formerly known as the Wind Rover. The early English name of the Swallowtail butterfly, *P. machaon*, was The Royal William. These popular names, whether in English or some other language, are often coupled with the scientific names but strictly have no place in zoological nomenclature. The earliest valid names date from the publication in 1758 of the tenth edition of Linnaeus's *Systema naturae*.

Linnaeus died in 1778, and in 1784 an English doctor named John Edward Smith bought his natural history collection, library and manuscripts. With several other naturalists he founded the Linnean Society in London and this has remained one of the most important societies in the world. Among its early members were such eminent scientists as Darwin and Alfred Russel Wallace, who lectured there and expounded the theory of evolution by natural selection. Linnaeus's collection is of primary importance for taxonomic studies of the Lepidoptera and other insects, since it contains numerous type specimens on which the original descriptions of species were based.

Towards the end of the eighteenth century Linnaeus's system of naming and classifying animals became more generally adopted and its influence was evident in the increasing number of works on the Lepidoptera. Printing methods improved and more books began to appear with colour illustrations. One of the finest and most important of these early works is the *Sammlung europäischer Schmetterlinge* by Jacob Hübner which appeared as a series of volumes from 1796 to 1823. Hübner, a most painstaking

lepidopterist and a miniaturist of rare talent, raised the illustration of Lepidoptera to a new level. His original hand-coloured plates are preserved in the entomological library of the the British Museum (Natural History). The mid nineteenth century saw the publication of G. A. W. Herrich-Schäffer's *Systematische Bearbeitung der Schmetterlinge von Europa*, 1843–1856, which enlarged upon Hübner's work and was also superbly illustrated.

A reference collection of butterflies and moths is essential for any serious taxonomic studies. Providing collecting is not indiscriminate and only a few selected specimens are taken, little harm is likely to be done. Undoubtedly the best and most interesting way to learn about Lepidoptera is to rear them if possible. Some helpful guidance on collecting, rearing and preserving specimens can be found in *Practical Entomology* by R. L. E. Ford. The one item of equipment which every collector must possess is a butterfly net. It is not known who first invented this piece of apparatus, but Bernard D'Abrera jokingly suggests in his book *Moths of Australia*: 'Both the Russians and the Chinese claim this as a national achievement. The Russians attribute the invention to their well-known scientist Alexander Netsky Popov (the word is an abbreviation of his second name), while the Chinese insist that the word was first used in reference to the function of the Great Wall to keep the peasantry within the exploitive range of their imperialist class. It is, however, well known that the 'net' was first invented by an Englishman who had his design stolen by an American who saw in it the chance of a million dollar success!'

Above left : An eighteenth-century drawing showing caterpillars and pupae with their respective adults.
Above : Plate No. 195 (dated 1779) from Cramer's rare work on exotic Lepidoptera. He named these species as follows : A—Papilio minos; B and C—Papilio hippo; and D and E—Papilio polynice.

The Adult or Imago

The adult or imago of an insect is its final stage of development. At this point, it is sexually mature and able to mate and reproduce its species. All species of Lepidotera pass through four distinct stages: the egg or ovum (embryo stage); the caterpillar or larva (growing stage); the chrysalis or pupa (transition stage); and finally the adult or imago (reproduction stage). This four-stage development in insects is known as complete metamorphosis and the Lepidoptera are accordingly classed as Endopterygota or Holometabola. Some orders of insects have partial or incomplete metamorphosis, that is to say they do not go through four distinct stages of development like the Lepidoptera. Instead in the immature stages they are generally nymphs which resemble the adults and they do not go through a pupal stage. These insects are classed as Exopterygota or Hemimetabola.

Insects with a complete metamorphosis have generally been much more successful than the others in adapting and spreading themselves in the environment. Lepidoptera are among the most successful of all, having developed scaled wings which give them mobility and also protection from their predators. The colouring and marking of the wings often camouflages the butterfly against its environment, sometimes by making it resemble an object or animal around it; sometimes the colouring acts as a warning to potential enemies. The power of flight and the exploitation of wing scales is found in all butterflies and moths, ranging from the pigmy Nepticulid moths with a wingspan of only a few millimetres to the giant Atlas moths and Birdwing butterflies with wingspans of up to 300 mm (11 ins).

The body of an adult lepidopteran is divided into three distinct regions: head, thorax and abdomen. The form and structure of these reflect the fact that the existence and survival of flying animals depends strongly on their ability to manoeuvre safely, to find food for their progeny, to avoid enemies and to find a mate. Much depends on the development of efficient visual and sensory organs of perception. In

Left: An exotic nymphalid butterfly with short, pointed tails on its hindwings.

the Lepidoptera the most important of these on the head are the antennae and the large, globular compound eyes. The head may also bear a pair of simple eyes or ocelli. Elaborate mouth-parts have been evolved from a primitive chewing type to form a specialized sucking tube known as the proboscis.

The compound eyes are usually very prominent and positioned to the side of the head below the antennae. They are a type of eye that has been developed to suit the particular requirements of highly mobile creatures which mostly need to be able to recognize shapes, patterns and colours from close-up. Thus a type of eye has been evolved which does not give the long-distance visual keenness of birds and mammals, which have in the main quite different requirements from insects, but an eye well suited for detecting movement close-up, a vital requirement for small creatures who have numerous predators. The eyes are referred to as compound because they register a compound image—not one complete image as in the human eye, but an image built up in a mosaic fashion from numerous little rod-like units called ommatidia. Each ommatidium has its own corneal lens and light-sensitive apparatus and an optic nerve, and transmits to the brain the image of one minute fraction of the eye's total field. Although the compound eyes do not have an iris diaphragm of the type found in the human eye, most of them, especially those of nocturnal species, can change the light sensitivity by expansion and contraction of the pigment in the retinal cells. This is similar in effect to the gradual adaptation of the human eye to changes of light intensity.

An important feature of the compound eye is its capacity for assessing distance. Human eyes can move so as to keep an object in focus when it gets very close, but the eyes of Lepidoptera are fixed and cannot move. Instead, as the insect approaches an object, the image on the retina gradually moves towards the inner part of each eye and the visual angle of the ommatidia becomes progressively less over the inner part of the eye. Perception in some species is poor but in others it is good. A number of Lepidoptera, especially moths, are able to see to a great extent by ultraviolet light which has wavelengths of less than 3900–4000 Ångström units, which is beyond the upper limit of the human eye vision. It has recently been discovered that this enables certain species to discern the particular patterns of flowers, and also to recognize their own species. On the other hand, many nocturnal moths are insensitive to red light, which is at the other end of the spectrum, and are not disturbed at night if illuminated by a torch with red glass.

In addition to the compound eyes, a pair of simple eyes or ocelli may be present on the cranium or upper part of the head. These are usually situated slightly above the compound eyes and are partially hidden by scales. Ocelli are made up of a corneal layer and a layer of sensory cells; there is no crystalline material between the two as in the compound eye. They are unable to form an image, but they are exquisitely sensitive to changes in light intensity, and it now appears certain that they influence the activity of the compound eyes in response to stimuli of this very particular type.

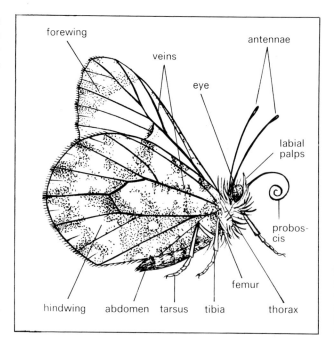

Left : A lateral view showing the main anatomical structures of the Orange-tip butterfly, Anthocharis cardamines *(Pieridae).*

Left : The feathery bipectinate antennae of an exotic moth. This type of antenna is particularly common in males.

The chief perceptors of smell are the antennae, which function as detectors, carrying sensilli that respond particularly to chemical stimuli. These are especially well developed in the male and are used in tracking down females of their species. They are segmented and flexible, being made up of anything from about seven to a hundred tiny segments. Their shape varies enormously from relatively simple and threadlike (filiform) to highly intricate plumose or feathery forms (pectinate), and the shaft may be either bare or covered in scales and sometimes in fine hairs (ciliate). The antennae show an almost infinite number of modifications, usually of a secondary sexual nature in the male. In some groups, especially phycitine moths, the basal segment is expanded and has a sinus or groove containing a special hair-pencil. In some of the primitive families the basal segment is dilated to form an elaborate eye-cap over the compound eye, while in other groups a long pecten of hair-like scales projects over the eyes. The longest antennae are found in the adeline micro-moths (Incurvariidae), where in the male they can be several times longer than the forewings. In contrast, the females of the primitive psychid moths often have degenerate and atrophied antennae. But whatever the shape and degree of development of the antennae, they are the site of sense organs, particularly touch receptors and chemoreceptors, and are vital to an active lepidopteran's existence.

The life-cycles of these creatures are closely associated with flowering plants and parts of their anatomy reflect this relationship. Their tongue which is known as a proboscis or haustellum, is of a sucking type suitable for drinking nectar. Only in the archaic Zeugloptera and some primitive forms in the Dacnonypha is the ancestral chewing or biting mandibulate mouth still found.

The spirally coiled proboscis found in most butterflies and moths is an exceedingly complicated structure. Its evolution follows a general degeneration of the mandibles and the compensatory development of other structures, particularly the galeae. The latter structures have become elongated, with the inner side of each grooved, and have developed

Below : A frontal view of a wasp-mimicking clearwing moth (Sesiidae). Note particularly the antennae composed of a series of segments.

hooks and spines which interlock and fasten both galeae together so as to form a single hollow tube. When the adult emerges from the pupa the two galeae are still separate and must be fitted together. Fluids are sucked up by muscular action of the pharynx. Each galea has an elastic strip of cuticle in its upper wall which serves as a stiffener and keeps the proboscis coiled. Uncoiling is effected by the contraction of muscles spread along the upper surface. In the very long proboscis of some butterflies, about one third of the way along from the head is a weak point where the proboscis can be bent almost at right angles when extended thus facilitating feeding at close quarters.

When the proboscis is not in use it is kept coiled up beneath the head and thorax, and in some of the larger butterflies, in which it is well developed, it looks rather like the hair spring of a watch. It varies greatly in length, attaining its maximum in some of the hawk-moths or sphingids where it can be more than twice the length of the body; this enables them to suck nectar from deep flowers, such as the convolvulus. In other groups, the proboscis may be rudimentary or absent, as for instance in some cossids, psychids and lasiocampids in which the adult does not feed. In some species the proboscis is short and stout; this feature enables the Death's head Hawk-moth, *Acherontia atropos* (Sphingidae), to penetrate the lids of the honey-cells in a beehive, or certain tropical noctuids to perforate the skins of citrus fruit and other soft fruits, thus making them agricultural pests.

Adjacent to the proboscis are two pairs of segmented tactile organs or palpi. One pair is known as the maxillary palpi, and these are usually small and inconspicuous, varying from 5-segmented and folded palpi in the primitive families to 1-segmented palpi in the advanced groups. The other pair is the labial palpi which is normally 3-segmented and often very long and prominent.

The thorax is the locomotory region and comprises three segments each of which bears a pair of legs. The segment nearest the head is the prothorax and is generally small. It has a pair of articulated lobe-like projections known as the patagia that form a collar behind the head. The patagia often have differently coloured scales from the head and thorax and can be useful in identification. The mesothorax and metathorax are more strongly developed and each carries a pair of wings and, overlapping the base of the forewing of each pair, a small articulated sclerite known as the tegula which is covered with scales and looks rather like a miniature epaulette. In some Notodontidae, Thaumetopoeidae, Lymantriidae, Arctiidae and Noctuidae, the metathorax bears a pair of tympanal organs. In certain other families—Geometridae, Pyralidae—similar organs are situated at the sides of the first abdominal segment. Tympanal organs are auditory and can pick up ultrasonic vibrations, such as those produced by bats, as well as those of lower frequencies.

The three pairs of thoracic legs—one pair on each segment—are usually long, slender and well developed. They are jointed and consist of the coxa which is articulated with the body, followed by the trochanter, then the femur and lastly the tarsus which is 5-segmented and has a pair of claws on the terminal segment. The tarsi of many Lepidoptera have highly sensitive chemoreceptors which can detect sugar and saline solutions with a greater efficiency than that of the human tongue. The front pair of legs may be extremely small, particularly in some of the nymphalid butterflies, and useless for walking, but often have brush-like scales for scraping leaves. In the Lepidoptera the coxae are characteristically set close together mid-ventrally. The tibiae of the middle and hind legs of some species carry groups of scent-bearing scales, while the front legs may carry a lamellate spine or spur on the inner side. This spur is known as the epiphysis and, together with the first segment of the tarsus, can be used to clean the antennae and the proboscis.

In the Lepidoptera, both upper and lower surfaces of the wings are characteristically covered with minute scales of various shapes and colours. The wing, an organ characteristic of the adult insect, develops as a lateral expansion of the junction between the dorsal and lateral plates of the segment concerned. It consists of two closely opposed layers between which runs a more or less complex supporting system of sclerotized tubes containing tracheae, nerves and blood. This skeletal network of tubes is referred to as the wing venation and is much used in the classification of families and genera.

The wings of Lepidoptera are generally roughly triangular in shape and have a large surface area in comparison with that of the body. In the male, especially in butterflies, the wings often have patches of specialized scales, usually thin and elongated and with a tuft of hair at the tip; these scales are connected to scent glands and are known as androconia.

Many families include at least a few species in which the wings are reduced and are useless for

Below: A microscopic view of the head of a lepidopteran. Note the segmented antennae, the large globular compound eyes, the three-segmented labial palpi, and the coiled proboscis.

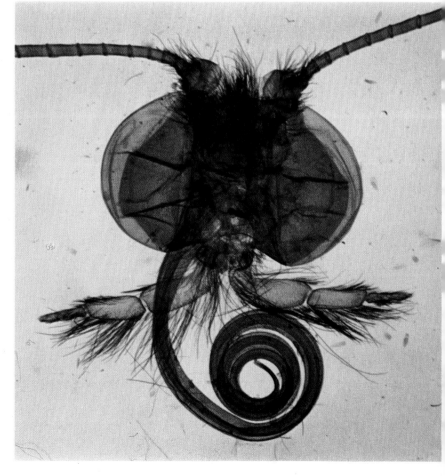

Above: The head of the Two-tailed Pasha, Charaxes jasius (Nymphalidae). The unrolled proboscis of the butterfly shows the two constituent galeae separated at the tip.

flying. This usually occurs in the female and is common in the Psychidae, the genus *Orgyia* (Lymantriidae) and various Geometridae. Adults having reduced wings are variously known as brachypterous, micropterous or subapterous, while those that do not possess wings at all are called apterous.

The wing venation of the Lepidoptera is of outstanding simplicity but nevertheless shows characteristics which can separate the different groups down to generic and even species level. The wings are traversed by two types of vein, the longitudinal veins which ramify or branch to varying degrees, and the cross-veins which divide the wing into areas known as cells. The longitudinal veins are: the costa, which runs along the leading edge of the wing and never branches; the subcosta a short distance behind it; the radius, which is generally thick and divides into several branches, sometimes forming an accessory cell with a chorda; the media, formed by three branches at most; the cubitus, also generally subdivided into several branches; and the anal veins, varying in number between different species and

groups. The media is often absent in its proximal part, and its branches, which are connected by cross-veins in the middle part of the wing, form the distal boundary of a structure typical of the Lepidoptera, the discoidal cell.

The forewing is approximately triangular in shape and has a humeral (basal angle) where the wing articulates with the thorax, an apical angle or apex at the forward outer corner, and an inner angle at the lower outer corner or tornus; these points define the edges of the wing, which are known as the costa (front edge), termen (outer margin) and dorsum (inner margin or back edge). The hindwing, which is also usually roughly triangular, is more rounded towards the apical margin, and shows marked morphological variation between groups.

Butterflies and moths fly by beating their front and hind wings in unison; the two pairs are generally linked by one of a number of types of coupling mechanism. One such mechanism, the so-called 'frenate' type of coupling is common among the Ditrysia and is formed by a frenulum consisting of a single composite bristle in the male and usually two or more separate bristles in the female, arising from the humeral angle of the hindwing. The frenulum engages with a retaining device called the retinaculum. This differs between the sexes, and in the male is in the form of a curved subcostal hook, while in the female it is more simple and usually consists of stiff bristles or scales situated near the base of the subcostal and, or, subcubital veins. In the Rhopalocera and a few ditrysian moths the frenulum is reduced or lost and an amplexiform type of coupling has evolved in which the basal area of the forewing substantially overlaps the expanded humeral lobe of the hindwing.

The flight of Lepidoptera varies in speed and power, depending on the shape and surface area of the wings and the rate of the wing beat. The number of beats per second varies from 5 to 12 complete beats in slow-flying butterflies to 50 to 70 per second in some fast-flying moths. This depends, among other things, on environmental conditions and particularly on the surrounding temperature. When the temperature is too low, the insects raise the temperature of their own muscles by rapidly vibrating their wings before taking off. It has been demonstrated that, in the Humming-bird Hawk-moth, *Macroglossum stellatarum* (Sphingidae), the beating of the wings raises body temperature by 2°C (3.5°F) per minute, and that in order to take off the temperature must exceed a particular threshold, varying from 26.5°C (79°F) to 36°C (97°F) in the Hawk-moths, and higher than 30°C (86°F) in other Lepidoptera such as vanessid butterflies. Other species achieve the same effect by exposing themselves to the rays of the sun, and can only take off when they have warmed themselves up sufficiently.

Flight speed is very variable among butterflies and moths; it is highest in the Hawk-moths which can reach a speed of 15 metres (16 yards) per second, or 54 kilometres (34 miles) per hour. The fastest species fly more or less in a straight line, while the slow fliers usually take an erratic zigzag course. This helps them to avoid or escape predators which by day are insectivorous birds in particular and by night are bats.

Butterflies are generally known above all for the colourful markings and patterns on their wings and bodies. These colours may be structural, pigmental, or a combination of both. Structural colours may be produced by light interference, and usually have a metallic or iridescent quality. They can always be distinguished from pigmentary colours by the fact that they disappear as soon as the structure that gives rise to them is damaged; as, for example, by crushing the coloured part. Structural coloration is highly developed among butterflies, whose wings are usually covered with minute scales. It is the parallel lamellae, the striations and granularities of these scales that give rise to the light-effects associated with a particular colour.

White pigments do not exist in nature. White coloration is due to the total reflection, in all directions, of the light falling onto a structure consisting of minute particles, or containing microscopic bubbles of air. When any of these particles is illuminated in an appropriate manner, it becomes the centre of an explosion of light which gives the illusion of whiteness. Thus, the brilliant patches of white on the underside of the hindwings of certain butterflies

are produced by a layer of air lying between the two layers of the wing membrane.

Iridescence is often found in blue, purple and green-coloured butterflies, and is produced by a system of thin layers or lamellae. The scales of these insects bear very fine alternating layers of chitin and air, which may lie parallel or perpendicular to the surface of the scale. When the wing is viewed from a very oblique angle, parallel lamellae give a red appearance and perpendicular ones appear violet.

Lamellae lying perpendicular to the surface of the scales and varying in thickness from 1 to 3 microns are used to produce eye-like marks, with different colours at the centre and the periphery. This is due to the fact that the lamellae are thicker at the periphery than at the centre, so that the play of light off them is different.

The process of breaking sunlight up into its component colours, as in a prism or a diffraction grating, is used by a number of insects. The parallel grooves and ridges used for the purpose need to be spaced out at distances of the order of the wave-

Below : A portion of the wing of a tropical American Morpho butterfly (Nymphalidae) shown at high magnification to demonstrate the scales. The metallic colours of these insects vary depending on the angle of incidence of the light.
Right : The magnificent colours of the large eye-spots which adorn the wings of the Peacock butterfly, Inachis io *(Nymphalidae).*
Below right : The Silk moth Graellsia isabellae *(Saturniidae).*

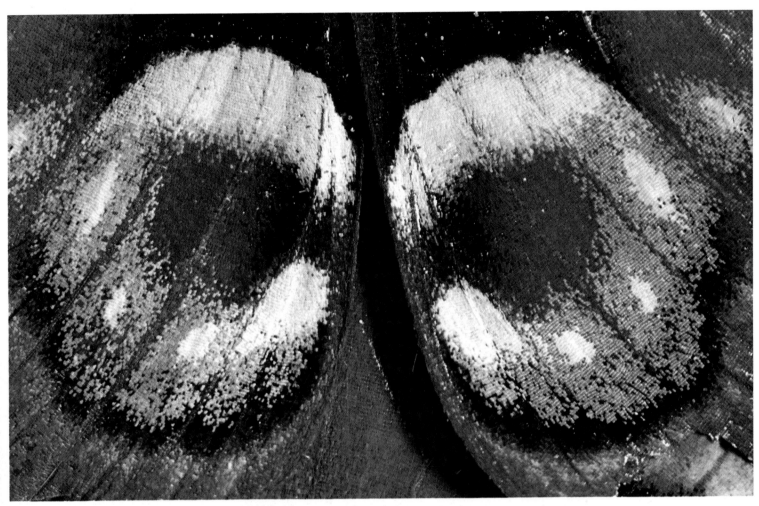

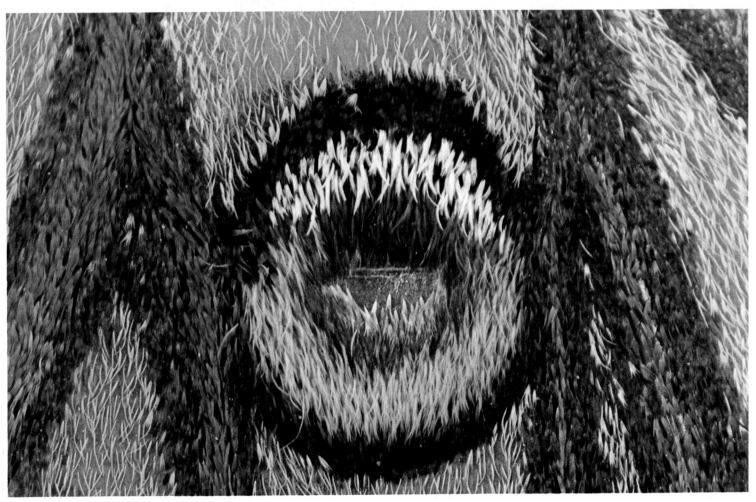

length of light. The scales of butterflies of the genus *Morpho*, with their splendid iridescent blues and metallic sheen, have been shown to carry up to 1400 striations per millimetre.

Most insects, however, have colours based on pigments, coloured compounds lying below the surface of the insect. There are very many different types of pigment, some of which are directly derived from ingested food materials and others are by-products of metabolism. Research on vanessid butterflies has shown that their red pigment is derived from chlorophyll. It is formed from the green pigment contained in the leaves that the caterpillar eats, which is gradually distributed in the haemolymph to all parts of the body. The green plant pigment is then converted into red animal pigment so completely that even the insect's excrement is coloured red.

There are certain cases in which pigmentary and structural coloration are combined. One example is the Paradise Birdwing, *Ornithoptera paradisea* (Papilionidae), whose wings are green and yellow. The scales have lamellae perpendicular to their surface, and the structural colour thus produced is blue. However, the body of the scale and the lamellae contain a yellow pigment, which interacts with the blue to produce green. A related species is blue on top and green below, since the scales on the upper surface of the wing do not contain yellow pigment.

The iridescent green scales of certain Lycaenidae carry a large number of small cells which reflect green light of varying intensity, depending on the amount of brown light transmitted through the transparent scale. Another type of butterfly is dark blue, with patches of pale blue or white on the wings. Close examination reveals that the structural colour is the same throughout, but it is heightened by a black pigment which is present in the areas that appear deeper blue; the black screen absorbs light-rays of other colours that would otherwise dilute the basic hue.

Adult Lepidoptera have an abdomen composed of 10 segments plus the anal segment or telson. Eight segments can be clearly identified in the male and seven in the female, the remaining terminal segments being highly modified to form the structures of the genitalia. In both sexes of most Lepidoptera the genitalic structures provide characters of great value for classification from suborder to species level. The terminology of the parts that make up the genital segments of the Lepidoptera is extremely complicated and not always uniform, which is largely a result of the extreme variability in the morphology of this region.

In the male, the ninth segment and part of the tenth form the tegumen, a hood-like or roughly triangular or trapezoidal plate that sometimes is extended laterally; sternum nine forms the U-shaped vinculum, which may carry a median process or

Left : A beautiful, exotic papilionid butterfly. Below : The Humming-bird Hawk-moth, Macroglossum stellatarum *(Sphingidae), a day-flier which is often seen darting around flowers in gardens in Europe.*

saccus on its anterior part, and a pair of clasping organs or valvae, laminar appendages of the most varied appearance. The median portion of the whole complex carries the penis or aedaegus, which has an evaginable portion or vesica composed of a membranous tube, which often carries spines or thorns on its surface. The tenth segment is much reduced in size, and typically consists of the uncus, a pointed distal extension of the tegumen, and the U-shaped gnathos, which together with the uncus forms a sclerotized ring enclosing the anal tube.

The female genital apparatus is more complex and reflects the evolutionary development within the higher categories of the Lepidoptera. There may be two genital openings, one for the reception of sperm at copulation and the other for egg-laying; or one opening may perform both functions. The former arrangement is known as ditrysian and is seen in most Heteroneura, suborder Ditrysia; the latter is known as monotrysian and is found in the archaic Zeugloptera and the primitive Dacnonypha and Monotrysia, except in the Hepialoidea where an

intermediate type is found known as exoporian. In the primitive monotrysian genitalia the external genital aperture or cloaca is situated in the area of the last two segments, at the posterior extremity of the abdomen. In the ditrysian genitalia the separate copulatory opening or ostium is situated on the eighth sternum. The intermediate exoporian type found in the Hepialoidea has a separate opening, but this is situated in the region of the last two segments close to the anal aperture. The female has a pair of ovipositor lobes or papillae anales on these segments. These may be soft, hairy lobes or rigid and sharply pointed. Sometimes they are developed into a long flexible ovipositor, as in the Leopard moth, *Zeuzera pyrina* (Cossidae), which is generally used for depositing eggs deep in the crevices of different kinds of bark.

The abdomen of Lepidoptera is almost always covered with hairs and scales. In the male, there is often a tuft of long hairs covering the copulatory apparatus; and in the females of various families (Lymantriidae, Lasiocampidae, Thaumetopoeidae)

Above: A many-plumed Moth, a strange member of the Microlepidoptera belonging to the genus Alucita *(Alucitidae). These moths have each of their wings split into six lobes or plumes. Right: An uncommon Asiatic butterfly; the pale colours of the undersurface of the wings contrast with the splendour of the upper surface.*

the tip of the abdomen has a huge clump of deciduous hairs which are used to cover the eggs as they are laid.

Anatomically the nervous system of adult Lepidoptera is no different, in any essential respect, from that of insects in general. Among sense-organs, however, we must single out one that is a characteristic feature of this order. This is the chaetosema, which is situated on the head and consists of a pair of papillae thickly covered with hairs. They lie behind the antennae and close to the compound eyes, and are connected via nerves to the brain. The precise function of this organ is not known but it is almost certainly sensory. It is present in Rhopalocera and in some families of Heterocera.

The digestive apparatus is divided, in both adult and larval stages, into three regions that have different embryonic origins. In the adult, the mouth-parts are of the simple biting type or an advanced sucking type in which the most conspicuous component is often the coiled proboscis. Food is sucked in via the pharynx, which is equipped with powerful dilator muscles that enable it to act like a suction-pump—

behind it lies a narrow oesophagus. In some primitive forms for example the Micropterigidae and Hepialidae, this shows a small local dilatation, while in higher forms it opens into a diverticulum, (a blind pouch), which is often bilobed, and into the alimentary canal.

The circulatory system includes a pulsating dorsal vessel that runs the length of the abdomen. It carries seven pairs of openings or ostia, each with a valve that allows the blood or haemolymph to enter but not to leave. There are also other accessory pulsatile centres whose distribution varies.

Finally respiration takes place through the spiracles on the abdomen which connect to a highly complex network of fine branching tubes or tracheae within the body. The Hawk-moths and other efficient fliers have so-called air-sacs, which are dilatations of the tracheae, which function both to lighten the body and to provide reserves of air. This system of breathing limits the size potential of insects because the diffusion of oxygen is only effective over extremely short distances.

Courtship, Mating and Reproduction

Animals which are capable of flight, and which may range over a wide area in search of food, are likely to have special problems when seeking a mate. An individual moth or butterfly species may not only be widely distributed but, particularly in the tropics, may fly in company with many other species similar to it in colour-pattern and habits. Even in temperate regions the problem of mate recognition is very real, especially in large nocturnal groups like the Pug moths, *Euphithecia* (Geometridae) where numerous apparently nearly identical species may inhabit a given area. Methods by which males and females attract each other include visual, olfactory (by smell), acoustic (by sound), and tactile techniques, as well as those involving a combination of infra-red radiation and scent. How these techniques are employed and combined depends on the powers of flight of the insect, its operating medium (air or ground – some female moths are wingless), the timing of mating activity and life-cycle, the insect's operating altitude and its geographical location. The luminescence produced by nocturnal Firefly beetles (Lampyridae) is a particularly exciting method of mate-location but has yet to be discovered among the moths, most of which are nocturnal and possible candidates as light-producers.

Moths are active mostly only at night and the females of the species have solved the problem of finding a partner by the use of attractive scents recognized by minute sensillae on the antennae of a potential mate. Female moths fly little, some not at all, but as virgins produce a specific sex-scent or sex pheromone, (derived from the Greek words *pherein*, to transfer, and *hormon*, to excite) which will eventually attract a searching male. A male will be attracted only by a female of its own species and may be actively repelled by female scents of other species.

Extremely small quantities of female scent can be detected by a male. A male Gypsy moth, *Lymantria dispar*, (Lymantriidae), will react to as little as one ten thousand millionth part of a gram of the female

Left: A pair of Map butterflies, Araschnia levana *(Nymphalidae), named after the design of the undersurface of the wings.*

Gypsy moth's sex scent, and will even attempt to mate with pieces of blotting-paper impregnated with a few drops of scent extracted from the sex glands at the tip of the female's abdomen. The sex scent of the female Gypsy moth can now be made in the laboratory, and as the commercially known Dispalure has been used in attempts to control this destructive species in North America. Traps have been baited with Dispalure and it has been sprayed over tracts of forest to confuse the males and reduce the number of successful matings.

When a virgin female moth has produced her sex pheromone, she remains at rest and awaits the attentions of a male. The males fly about, apparently at random, until they encounter the scent of a female. Whether males are able to follow scent trails as long as a kilometre or more is doubtful, but once having come within a few metres of a 'calling' female the appeal of her scent is irresistible to the male. Female sex scents may be produced only at certain times of the night, although in rare cases such as in the partly day-flying Promethea moth, *Callosamia promethea* (Saturniidae), they are produced in the day. The European Angle Shades moth, *Phlogophora meticulosa* (Noctuidae), for example, is receptive to males only between 1 a.m. and dawn, at which times it extrudes a bilobed scent-producing gland at the end of its abdomen while simultaneously vibrating its wings (probably to aid scent dispersal).

After the male has been attracted to his potential mate, he produces a scent and often performs a series of special courtship movements comparable in complexity to those of birds. The male's sex scents, which are manufactured by glandular tissue within the body, have to be transferred rapidly to the outside just before mating. This is achieved by means of often quite large scent brushes or hair pencils, which are associated with the scent glands and at the right moment are opened up or everted and effectively scatter the scent molecules. Many of these scent-distributing organs resemble small dandelion (*Taraxacum*), seed-heads; others are simply patches of specialized scent-impregnated scales called anroconia. Male sex pheromones are often referred to as

aphrodisiacs, but this may not be a universally appropriate term as the female is frequently sedated or made passive rather than excited by the male scents.

An attractive theory proposed by E. R. Laithwaite suggests that the feathery antennae of many male moths act like aerials and receive waves of radiation from the far-infra-red part of the electromagnetic spectrum. Later experiments have so far failed to confirm his theory, but one has only to consider the ease with which light, another part of the spectrum, is perceived by an eye (whether human or insect), to realize that the perception of radiation is a common-place phenomenon. P. S. Callahan considers that moth scents luminesce in various infra-red colours and that this luminescence is visible to moths by means of minute sensillae on the surface of their antennae. It seems therefore that radiation and scent have an integral part to play in communication between nocturnal moths.

Sound provides another possible method of nocturnal signalling. Most moths do respond to sound and many have ears, but their response seems to be related to bat cries, not to moth sounds. Many Tiger-moths (Arctiidae), of both sexes produce sound at night as a warning to bats of their noxious qualities. In fact, moths do not seem to use sound as a sexual attractant in the same way as certain mosquitoes, whose females attract males by the particular pitch of sound produced by the female wing beats. However, sound is probably used territorially by the European and Asian Tiger-moth, Setina aurita (Arctiidae), which has larger sound-producing tymbal organs in the male than in the female. Day-flying males of the Australian Whistling moth, Hecatesia fenestrata (Agaristidae), also produce sound by repeated contact of a cupped area on the forewings which is distorted at the top of each wing-beat. Other moths use a file and scraper method, for example the Australian agaristid, Platagarista tetra-pleura, and the Madagascan agaristid, Musurgina laeta, whose hind legs rub against ridged areas on the hindwing and forewing respectively. Perhaps unique, and remarkably similar to human voice

production, is the method used by the Death's Head Hawk-moth, Acherontia atropos (Sphingidae). This species sucks or blows air through its pharynx, the air passing over a projection called the epipharynx producing a pulsed air-flow or sound. The short proboscis probably acts as an amplifier. Various functions for the Death's Head moth's squeaking or hissing sound have been suggested, including the possibility that hive-bees mistake this sound for the pre-swarming sound made by a queen bee—a useful piece of deception for a species of moth said to visit old-fashioned bee hives to feed on honey.

The sound-receiving organs or ears of most moths are found on the sides of the body, either at the rear part of the thorax as in the Owlet moths (Noctuidae), and their allies, or at the front part of the abdomen as in Geometridae and associated families. Noctuids and geometrids have a thin ear-drum or tympanic membrane covering an air-filled cavity surrounded by a tough chitinous ring. A tympanic nerve trans-mits sound messages from the ear to the brain. The American biologists K. D. Roeder and A. E. Treat found that noctuid moth ears responded to sound frequencies from three kilocycles per second, (within the human sonic range), to 100 kilocycles per second, but that they were most effective at frequencies in the middle of this range corresponding in pitch to sounds emitted by hunting bats. (Human ears can hear sounds as low as 30 cycles per second but are deaf to sounds higher in pitch than about 20 kc/s.) Some Hawk-moths (Sphingidae), hear through two ap-pendages of the head called palps and pilifers which when in contact are receptive to ultrasonic fre-quencies. The tropical American Passion-flower butterfly Heliconius erato (Heliconiinae), has its ear at the base of the hind wing in the form of an air-filled sac, while the Cracker butterfly, Ageronia feronia (Nymphalinae), also of the New World tropics, has its sac-like ear at the base of its forewing.

Sound production and reception is not confined to Lepidoptera; many other groups of insects scrape, tap, vibrate or produce a variety of sounds by other means, many of which appear to be sexual in function. Tappers include beetles of the genera Anobium and

Xestobium (Anobiidae), which strike their heads against the wood in which they live, and booklice (Psocoptera), which tap the ground with their abdomen. The loudest noise is probably made by cicadas (Homoptera), which use a vibrating membrane mechanism similar to that of arctiid moths. Loudness is an especially useful attribute for males of the North American cicada genus *Magicicada*; the noisier they are the more successful they are in attracting females. Grasshoppers, Bushcrickets (Katydids) and their allies (Orthoptera) are doubtless the principal producers of sound in the world of insects. They make their sounds mostly by scraping or frictional methods, usually involving the rubbing together of two different parts of the body; antenna and mandibles, leg and abdomen, leg and forewing, the two forewings, and other combinations. Specialized sound receptors are the rule in species of Orthoptera which may have the ears on the abdomen, thorax or the legs. Cicadas and some Plant-bugs (Heteroptera) have their ears on the thorax. Less obvious sound receptors are the numerous minute hair-like sensillae scattered over the surface of the head and body of many insects.

Female butterflies and probably some brightly coloured day-flying moths like uraniids and castniids rely, at long range, primarily on sight as the means by which they are recognized by their males. Colours, patterns and shapes are important, as is movement. Males of the Australian Ulysses or Blue Mountain butterfly, *Papilio ulysses* (Papilionidae), can be attracted to blue lures from over 30 metres (33 yards) away, but are attracted more effectively if the lure is fluttering in the wind. Differences in colour-patterns between the two sexes of a species are advantageous at this stage of courtship as they let males concentrate on fruitful heterosexual pursuits.

Many male butterflies appear to search in a random way for females, but it seems likely that the special methods adopted by some species may be more generally used than has been proved at the present time. Congregating at flowers, where both males and females are likely to be feeding together, provides one method of restricting a male's hunting ground.

Below : A pair of copulating Plain Tiger butterflies, Danaus chrysippus *(Danaidae). Ranging from the Canary Islands, the Mediterranean and Africa to Australia and New Zealand, this butterfly is unpalatable to predators because it contains heart poisons derived from the milkweed plant on which the caterpillar feeds.*

Hill-topping is a common phenomenon in the tropics. At certain times of day males are restricted to the tops of hills, to which virgin females make their way singly and are courted by several males. Skippers (Hesperiidae), Swallowtails and their allies (Papilionidae), as well as Blues and Hairstreaks (Lycaenidae), are all known to hill-top. Perching by males on a particular branch, leaf, or rock and awaiting passing females is common in the mostly tropical butterfly family Riodinidae but is by no means confined to this group. A perching male will chase away butterflies of other species, whether male or female, other insects and even small birds, but will begin to court a virgin female of its own species as soon as one arrives in its territory. Patrolling male butterflies fly over a larger area than perching males but restrict themselves to a particular woodland glade, river bank, or hill-top. Nymphalines and papilionids are some of the most commonly seen patrollers.

The sex scents of most male butterflies, like those of moths, operate during courtship displays and mating, the female demands proof of the male's species identity in the form of a scent before mating can take place. Specific sex scents are especially important to those groups of mimetic butterflies, such as Passion-flower butterflies (Heliconiinae), in which many groups of species contain several almost identically coloured species. In these butterflies wasteful non-productive cross-mating is avoided by the use of male scents which are peculiar to one species and act as a sexual signal, whether aphrodisiac or sedative to the female, but will have little or no sexual effect on the female of another species. Time may also be an important factor in mating; a species may have rather precise times of day during which females are receptive to the males and males are simultaneously looking for females. Different timings for sexual activity would tend to ensure that males of closely related species did not waste energy attempting to mate with females of the wrong species, and in the absence of other overriding factors selection could be expected to produce timing differences of this type. Timing could be affected also by predation pressures in that butterflies engaging in conspicuous aerial courtship displays will be at less risk if these take place at times of day when insectivorous birds are least active. It is conceivable that humidity, which often varies throughout the day, may be another factor influencing mating times, since humidity is known to affect the efficiency of

Left : Eggs of the Map butterfly, Araschnia levana (Nymphalidae), which are laid in characteristic hanging columns on the underside of a nettle leaf.
Below : Eggs of the Large White butterfly Pieris brassicae (Pieridae), the yellow and black caterpillar of which skeletonizes cabbage leaves.
Below right : Eggs of the Scalloped Hazel moth, Gonodontis bidentata (Geometridae), seen at high magnification.

man-made aerials and may also affect the operation of an insect's antennal sensillae. Temperature, especially in temperate regions which may have low early morning temperatures even in summer, is also important to many butterflies which cannot fly at low temperatures. The circumpolar Small Apollo butterfly, *Parnassius phoebus* (Papilionidae), will not fly below 13°C (55°F), while overwintering Monarch butterflies, *Danaus plexippus* (Danaidae), cannot be induced to fly below 10°C (50°F). Temperature-dependent alpine butterflies are mostly opportunists who make use of whatever warm weather is available and are capable of mating after an abbreviated courtship at any time of day.

The sequence of events leading up to mating has been described in detail for the Postman butterfly, *Heliconius erato* (Nymphalidae), of tropical and subtropical America. Males of this species are sexually most active at about 9 a.m. and are then especially attracted to the colour red. At this time of day they will fly around in groups round bushes and trees bearing red flowers, whereas earlier on in the day yellow flowers will have attracted them. Females have corresponding colour preferences at the same time of day as the male but do not congregate in groups; when a virgin female enters a group of males she will be courted immediately. A male and female will fly over and under one another, during which time the male sex scent is recognized by the female, who is urged downwards until at rest. The male continues to hover above and in front of the female, their antennae almost in contact, and then settles alongside the female, moves backwards a little and twists his abdomen through 180° until the claspers of his genitalia can grasp the female genitalia; he then moves so that the mating pair is end to end. Sperm is then transferred to the female in a gelatinous spermatophore via the penis of the male which is held firmly inside the female by an inflated vesica extruded from the end of the penis. The Postman butterfly always follows the same sequence of cues during courtship: firstly colour, then tumbling flight, smell, hovering flight and finally a sideways approach just before copulation takes place. If one of these links is missing the chances of successful mating are slight.

Courtship displays in other butterflies have been studied in the United States, especially that of the danaine Queen butterfly, *Danaus gilippus* which is found in the southern United States and tropical America. Like the Postman, the Queen has a set courtship routine which must be rigidly followed by both sexes. The male chases the female, overtakes her from above, brushes her antennae with the scent-brushes which extrude from the tip of his abdomen and eventually induces the female to alight. The male then continues to hover above the female, his scent-brushes still in action, until the female becomes suitably submissive: he then alights alongside the female and copulation takes place. Post-nuptial flights usually occur with the male hauling the female backwards through the air. Laboratory experiments have shown that males from which the scent-brushes have been removed will display in the usual way but are not accepted by females; females whose antennae have been made non-receptive do not react to male scent and refuse to mate. The male Monarch butterfly, *Danaus plexippus* uses its scent-brushes in a

similar way but it does so during a characteristic spiral flight. Some of the male sex scents of danaine butterflies are distinctive enough to be recognizable by man and are generally described as unpleasant; but one is supposedly reminiscent of frangipani flowers and another of vanilla.

Many South American ithomiine and danaine butterflies are known to extract the chemicals needed to produce their male sex scents from plants such as heliotrope (*Heliotropium*) and eupatorium (*Eupatorium*) which contain pyrrolizidine alkaloids. The butterflies either suck large quantities of surface water-droplets from the leaves and stems, or if the plant is dry, they will regurgitate a drop of fluid onto the plant, stir this drop with their proboscis and then suck it up again. The reason why some Tiger-moths and ctenuchid moths also visit these plants is not clear, but as females of some species are as frequent visitors as the males, the extraction of alkaloids for defence is a probability. This is a distinct possibility as pyrrolizidine alkaloids are known to be chemically partly responsible for the unpalatability of the Garden Tiger-moth, *Arctia caja*.

The range of colours used by butterflies as signals for species recognition extends from the red end of the visual spectrum into the ultraviolet regions in which man sees nothing. Many species of butterflies have distinctive patterns or hues of ultraviolet which are unseen by man (except when they are revealed by ultraviolet photography) but are readily recognized by the butterflies. For example, the male Small White butterfly, *Pieris rapae* (Pieridae), which is found in the temperate regions of America, Europe, Asia and Australia, is attracted to the particular ultraviolet hue of the underside of the female's hindwing. Birds, which are important enemies of butterflies, are also probably blind to the ultraviolet patterns by which some butterflies recognize one another. The vivid blues which appear so brilliant to us and are found in many butterflies, for example in Blues (Lycaenidae), and Morphos (Morphinae), probably do not attract the attention of birds so strongly, and can therefore also be used by the butterflies as recognition colours without blatantly advertising their presence to insectivorous birds. Reds, yellows and oranges, which birds can see acutely, are found in many butterflies and are important in species recognition between the sexes, but these colours are often restricted to the upper surface of the wings and are hidden from view when the butterfly is at rest.

The colours of Lepidoptera are produced by light-reflecting substances called pigments, by surface structure effects, or by a combination of the two. The brilliant iridescent colours of *Morpho* butterflies, Hairstreaks (Lycaenidae), and uraniid moths (Uraniidae), are all structural colours, but there is some doubt as to whether or not they are interference colours. E. R. Laithwaite believes that interference is not responsible for most structural colours in Lepidoptera and that some other factor is responsible. He has likened the colours of Morphos and uraniid moths (and those of aniline dyes), to those of metals, and has used the term metallic to describe them. Tests to show if a colour is structural or not can be made by viewing it from different angles (pigments remain the same colour whatever the angle of view, unlike structural colours), or by

wetting the surface with a wetting agent like ether which will temporarily eliminate any structural colours. Whiteness is also a structural effect in Lepidoptera in which the scales refract and reflect light of all wavelengths to produce white light, similarly snowflakes and white powders produce the same result. Pierid butterflies are white for this reason, curiously not because they contain a white pigment, leucopterin, which can be dissolved out of their wings without affecting their whiteness. Olive-green coloration is a special case in that it is produced in many moths and some butterflies, for example the Bath White, *Pontia daplidice* (Pieridae), by the juxtaposition of brown and yellow scales, not by an olive-green pigment.

Once mating has taken place the eggs which have been manufactured in the paired ovaries of the female pass down the oviduct, where they are coated with a protective shell. Sperm received from the male and stored in a membranous sac, the spermatheca, is transferred to the vagina where a single sperm will penetrate each egg through a minute pore or group of pores called the micropyle. Most species generally lay their eggs very soon after fertilization has taken place.

During fertilization the nuclei of the sperm and egg have united to form a zygote which in essence is a unicellular butterfly containing all the genetic material needed to reproduce the species. The zygote divides into two identical parts, each with a full complement of chromosomes. Millions more cell divisions then take place and eventually produce a minute caterpillar whose first task before emerging from the egg is to eat its way out through the egg-shell.

The sex of a moth or butterfly has been determined at a very early stage and is dependent on a particular pair of chromosomes in the nuclei of its cells. Chromosomes always occur in pairs, which are identical except for the sex chromosomes. These have two unequal parts, called X and Y, in female Lepidoptera. Male sex chromosomes have two equal parts called X and X. Sex cells (sperm or eggs) produced respectively in the testes or ovaries result from the division of cells having a normal complement of chromosomes so that each sex cell contains one half of each pair of chromosomes. Sperm will have only the X parts of the sex chromosome, whereas eggs will have either X parts or Y parts. Eggs will always be fertilized by X sperm so that equal numbers of XX (male) and XY (female) offspring will result when the two parts of the sex chromosomes reunite (see diagram).

Left: The typical and the dark or melanic forms of the Peppered moth, Biston betularia *(Geometridae). The melanic form (carbonaria) is apparently protected against predators in industrial areas, where the tree bark is blackened by atmospheric pollution.*

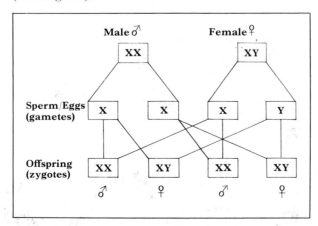

Only Lepidoptera, Caddis flies (Trichoptera) and birds have XY chromosomes in the female. The XY chromosome of man and other animals is found in the male.

Gynandromorphs are specimens which share male and female characters. These are often easily recognizable as such in Lepidoptera where there are obvious colour-pattern differences between the sexes. A gynandromorph which is male on one side and female on the other is fairly easy to spot; others may be male in front and female behind or have male and female characters distributed in a mosaic pattern over the whole of the insect. Gynandromorphs are genetic in origin and can result from an egg which has two nuclei, one having the X part of a sex chromosome, the other having the Y part, or perhaps the X factor has been accidentally eliminated in the egg during early cell division. Intersexes are less common and result from a change from one sex to another after development has been taking place for some time; they are dependent on a change in balance in the genes affecting sexual characteristics.

Parthenogenesis—reproduction by means of an unfertilized ovum—occurs in a few species of moths. Virgin females lay unfertilized eggs which develop normally to produce caterpillars and, eventually, fertile adults. The micropsychid, *Solenobia triquetrella,* for example, may produce only females for several generations. One subspecies of this moth has the normal number of chromosomes, the other has a double complement.

In some aphids (Homoptera) a sexual generation alternates with a parthenogenetic generation, a sequence of reproduction known as heterogamy. One apparent advantage of parthenogenetic reproduction is that mate location problems do not exist.

Polymorphism, a phenomenon in which one finds genetically controlled colour-forms of one or both sexes within one species, is common. The numerous female forms of the African Mocker Swallowtail, *Papilio dardanus* (Papilionidae), are a good example. Polymorphism is found commonly in mimetic Lepidoptera like *P. dardanus,* and the Passion-flower butterflies, (Heliconiinae), but is not uncommon elsewhere; for example, in the Silver-washed Fritillary, *Argynnis paphia* (Nymphalinae), a temperate European and Asian butterfly which has two female colour-forms, a brown form and a greenish-grey form.

Other well-documented Old World examples are the Mottled Beauty moth, *Alcis repandata* (Geometridae), which is dimorphic in western Britain (but not in the east), and the Wood Tiger, *Parasemia plantaginis* (Arctiidae), a brightly coloured species whose hind wings vary from white through yellow and orange to red, which has a white-winged form in northern latitudes and at high altitudes in Britain.

The confusion in names resulting from the existence of polymorphism and other forms of variation within a species is widespread—a species may have been described as new several times by different authors or, by the same author, and sometimes in different genera or even different families of Lepidoptera. Only when the offspring of a single female have revealed the potential genetic variability of the species has it been possible to associate two very dissimilar moths or butterflies as members of a single species. Who at first glance would assume

that a male and female *Papilio dardanus* belonged to the same species of butterfly? On the other hand, polymorphism is not invariably an instantly recognized phenomenon as instanced by the polymorphic blood-groups of humans.

The reasons for the evolution of different forms of a species differ from species to species; in *P. dardanus*, which mimics several different unpalatable butterflies, polymorphism enables it to increase its population without putting itself at risk by becoming too numerous in relation to its models, while in cryptically patterned species it is apparently advantageous to adopt more than one colour-pattern so that foraging birds, having 'got their eye in' with one pattern, may well ignore differently patterned though equally edible moths of the same species (an idea first put forward by the ecologist L. Tinbergen). Each of the two extreme colour-forms of the Peppered moth are less at risk to birds in different parts of its range which includes polluted urban areas, where sooty-coloured moths are best camouflaged, and rural localities where lichen-coloured moths are less easily distin-

guished from their background. Very dark or black (melanic) forms and other unusual specimens occur regularly in other species too, but their occurrence does not constitute polymorphism which involves a balance between two or more forms of a species and the various and varying selective pressures acting on the species. Polymorphic advantage to a species can be less obvious; for example, in humans there is a definite correlation between the different polymorphic blood-groups and resistance to certain diseases.

There are two colour-forms in the Silver-washed Fritillary, *A. paphia* the brown and the greenish-grey. If a predatory bird discovers that the brown form provides a highly acceptable meal it may then tend to choose another brown specimen so that greenish-grey specimens will be less at risk.

Variation of a genetic nature often occurs between populations of a species separated by a geographical barrier such as a mountain range. If the resultant differences between populations are large enough, interbreeding between them may be inhibited or impossible and a new species will evolve.

From Egg to Adult

All butterflies and moths begin life as an egg or ovum —this is the first of the four stages through which a lepidopteran passes during its life-cycle. The eggs which are laid by the adult are deposited on or near to the foodplant which will sustain the larvae, and are left untended. In a few rare cases the female is larviparous and the eggs are retained within her abdomen until they hatch.

The egg is soft when first laid and assumes its definitive form and colour as the shell dries and hardens. The shape is usually characteristic for family or generic groups, and varies from slender spindle-shaped eggs which stand erect to flattened plate-like eggs which lie flat on the surface; between these extremes come a variety of globular, pear-shaped, barrel-shaped and elongate forms.

At the moment of laying, the germinal cell is enclosed in two coverings: an external shell known as the chorion, and an internal one, the vitelline membrane. The chorion, which is secreted by the ovary of the laying female, is usually rigid and is imperme-able to most substances, though not completely impermeable to water; its inner surface may be coated with a wax to prevent desiccation and retain the moisture content of the yolk on which the developing embryo feeds. When the chorion is thin the diffusion of gases and therefore the process of respiration can take place over its whole surface. Where this is not the case, gas exchange takes place through pores known as pseudomicropyles, and also through an important opening known as the micro-pyle. This latter is an aperture in the chorion through which the male sperm can penetrate to fertilize the egg, and is usually situated at the summit of the dome or at the highest point, but in flat eggs may be at the side. The vitelline membrane lying immediately beneath the chorion forms the outermost layer of the yolk and is very thin and delicate.

Left: The large, stunning caterpillar of the Death's-head Hawk-moth, Acherontia atropos *(Sphingidae); this is a migratory species which is widely distributed in Europe and Africa. A typical feature of the larvae of Sphingid moths is the curved dorsal horn at the tail end.*

The outer surface of the chorion of some eggs is smooth and shiny, but many have raised ribbing or delicate filigree ornamentation or sculpturing, often in a pattern characteristic of the species. The micro-pyle may be in a simple depression in the surface or surrounded by a rosette-like pattern. Some of the smallest eggs are those of the pigmy nepticulid moths, with a diameter of no more than 0.2 mm (.0078 ins), while the largest of the Macrolepidoptera may be at least 3 mm (0.1 ins) in diameter. But the size of the egg is not always proportional to that of the adult; for example, the egg of the Emperor moth, *Saturnia pyri* (Saturniidae), one of the largest of the European moths, is only 2 mm (0.078 ins) across and is considerably smaller than those of other species of much smaller moths, such as the Oak Eggar, *Lasiocampa quercus* (Lasiocampidae).

The eggs may be of various colours and either unicolorous or patterned. They are commonly subdued shades of green, yellow or brown, but there are many exceptions, as for example those of the Blood-vein moth, *Cyclophora punctaria* (Geometri-dae), which are brilliant red, and those of some lycae-nid butterflies which are bright emerald green; while the eggs of the lymantriid moths of the genus *Orgyia* are black and white and those of the Lappet moth, *Gastropacha quercifolia*, are white marked with thick black curves and spots. In general, however, egg colour shows less variation than egg shape.

Some Lepidoptera lay their eggs in flight, dropping them at random, as for instance the hepialid moths, satyrid butterflies and the Silver-y moth, *Plusia gamma* (Noctuidae). A great many deposit them singly or in batches on a leaf of the foodplant, or in regular rows, as with the processionary moths (Thaumetopoeidae). Some pile the eggs haphazardly one on top of the other, like the Leopard moth, *Zeuzera pyrina* (Cossidae), while others are much more precise, such as the European Map butterfly, *Araschnia levana* (Nymphalidae), which lays its eggs in long pendant strings hanging from the underside of a nettle leaf, and the Lackey moth, *Malacosoma neustria* (Lasiocampidae) which deposits its large eggs in a tight bracclet around a twig. Many of the

leaf-roller moths (Tortricidae), which have flattened eggs, deposit them in overlapping, tile-like rows, which the female sticks to the substrate with a secretion from special glands. This secretion, which dries rapidly on exposure to the air, may also form a protective seal over the batch of eggs, as in the Satin moth, *Leucoma salicis* (Lymantriidae).

Often the eggs are protected with a layer of hair-like scales brushed off from the tip of the female's abdomen. Females equipped with an extensile ovipositor insert their eggs in the crevices of bark, stems, or into flowers and seeds, and sometimes actually pierce the epidermis. In the laboratory at least, the females of primitive micropterigid moths use their long ovipositors to lay small batches of eggs underneath stones placed in the breeding containers.

The total number of eggs laid varies greatly from species to species, but usually does not exceed two or three hundred, though in some species such as the large hepialid moths of Australia the total may be several thousand. As a rule, even if the female has mated, a considerable proportion of her eggs will not

have been fertilized. Food for the embryo is contained inside the egg in the form of yolk. As this becomes depleted the developing embryo can be seen through the shell curled up inside, its head appearing as a dark spot. With most externally feeding species the larva enters the outside world by gnawing a hole through the top of the shell. In leaf-mining species, such as the nepticulid moths, the larva gnaws through the bottom of the shell and burrows directly into the leaf. In some species the egg shell forms the first meal and is devoured entirely; it apparently contains some vital nutrients since larvae which are deprived of their egg shells at this stage have generally died.

One of the most important phenomena in insects is the ability to interrupt or arrest growth in order to overcome adverse environmental factors such as heat, cold, or short winter days. This phenomenon is known as diapause, and is a dormant state in which metabolism virtually ceases and the insect can survive over an extended period without food. Although usually induced by the onset of adverse conditions, diapause is obligatory in some species and will occur even if conditions are found to be optimal for continued development.

A classic example of obligatory egg diapause is found in the Cultivated Silk moth, *Bombyx mori*, whose eggs laid in the autumn will not hatch within 10 days or so like those laid earlier in the year, even if kept indoors in the warm, but will go into diapause and remain dormant until the following spring and then only hatch if they have been exposed to near-freezing temperatures. In some species the insect can be tricked and the diapause broken by placing the eggs (or larvae or pupae) first in a refrigerator and then in a warm room (an airing cupboard is ideal).

Insects can detect the approach of adverse conditions by subtle changes in their foodplant brought on by the advancing season, by a fall in the mean daily temperature and particularly by the shortening of the days as autumn approaches. For instance, the caterpillars of the Large White, *Pieris brassicae* (Pieridae), which feed on cabbages require more than 12 hours of daylight for the adult to develop, and emerge that season. If the day-length becomes shorter, any caterpillars feeding will continue until fully grown and will pupate normally, but their pupae will enter diapause until the following spring.

Different species are conditioned to diapause at different stages of their development. In some of the Tortricidae the larva hatches in the autumn and almost immediately constructs a tiny silken shelter or hibernaculum, sometimes without eating more than its egg shell, while in other tortricids, including the Codling moth, *Cydia pomonella*, the larva becomes fully grown and constructs its cocoon in the autumn but remains torpid inside until the following spring before developing into a pupa. Diapause may even occur in the adult, as for instance when the development of the ovaries is arrested. It differs essentially from hibernation and aestivation in that it is a condition where development is temporarily suspended. Hibernation is to pass the winter in a dormant state, as do the adults of the Peacock butterfly, *Inachis io*, and certain other vanessids; while aestivation is a state of dormancy during periods of summer heat or drought, as happens with the

Below : The caterpillar of the Spurge Hawk-moth, Hyles euphorbiae (Sphingidae), which feeds on spurges (Euphorbia).
Right : The cocoon of the Cultivated Silk moth, Bombyx mori (Bombycidae). The caterpillar is the well-known silkworm which was introduced into Europe around the year 550 by monks who smuggled both the moth's eggs, and the seeds of the mulberry on which the caterpillar lives, from China to Constantinople.

Australian Bogong moth, *Agrotis infusa* (Noctuidae), which retires to cool caves in the hot season.

The lepidopterous larva, or caterpillar as it is commonly called, is typically long and cylindrical, and is made up of 13 segments. The first is the hardened, more or less rounded (or more rarely flattened) head capsule, which is always well developed and distinct from the remainder of the body. This is followed by the three thoracic segments, known as the prothorax, mesothorax and metathorax respectively. Each thoracic segment typically bears a pair of segmented legs with a claw at the tip—these are the true legs of the insect which appear in the adult. The remaining nine segments form the abdomen, which is usually soft and flexible and generally bears five pairs of fleshy false legs or prolegs on segments 3 to 6, and 10, the latter pair being known as anal claspers. Many larvae appear to be quite smooth, but if examined closely will be seen to have a number of very fine hairs or setae, often arising from tiny circular spots. These are very important in classification and are known as primary setae. Often these setae are hidden among additional hairs, known as secondary setae, which can be very dense and are sometimes grouped in tufts on various kinds of tubercles or are developed into spines or some other ornamentation.

Although the head is very prominent in most larvae, in some species, usually those mining in leaves and stems, it can be retracted out of sight within the prothorax. It does not bear a pair of compound eyes as does the adult, but instead has a number of lateral ocelli or stemmata, which have the appearance of small rounded transparent lenses. There are usually six ocelli on each side, five of which are arranged in a semicircle while the sixth lies in front of them. Each ocellus has a lens and a receptive part or retina. Although known as simple eyes the ocelli cannot reproduce an image but probably only detect the difference between light and dark or the shadowy movement of an approaching predator.

Two antennae are also recognizable on the head of the caterpillar. They lie a little in front of and below the ocelli and are usually short, 3-segmented and bear several sensilla. At rest, the antennae can be retracted into cavities at the sides of the head; but the sensitive tip of each antenna remains exposed.

The caterpillar's mouthparts are typically of the chewing type. From above downwards, there is a labrum, a pair of mandibles, a pair of maxillae and a labium. The labrum (or upper lip) carries six pairs of bristles and a large number of sense organs.

The mandibles are quite tough, and well adapted to taking up and grinding the food. They may carry arrangements of crests or sharp points, which vary depending on the diet of the particular species. The lower lip arises from the partial fusion of the first maxillae with the labium, or lower lip proper. The labium, in its turn, arises from the midline fusion of the second maxillae. On it, the spinneret may be recognized: this is an elongated tube from which silk may be extruded.

In some mining larvae, the thoracic legs are much reduced in size or absent, and the first thoracic segment is larger than the others, so that the head can be withdrawn into it.

The soft, fleshy prolegs of the typical caterpillar are classified as ventral or anal, depending on their position. They are locomotory in function, and can retract and be used for gripping so that the larva remains securely anchored in one position, as anyone who has picked up one of the larger caterpillars of the hawk-moths and silk moths will have discovered. The proleg is generally broad and supplied with powerful muscle-bundles. Its foot part, or planta, is retractable and has a number of hooks or crotchets. The arrangement and number of these crotchets varies between the different family groups and is useful in classification.

The sides of the body, but not of the head, carry the spiracles or stigmata which are external breathing orifices leading directly to the tracheae which make up the respiratory system. There is a pair on the prothorax, none on the mesothorax and metathorax, and a pair on each of the first eight abdominal segments. The spiracles are usually easy to see as they have a peripheral rim or peritreme which is a different colour from the rest of the body.

A few caterpillars are aquatic, spending the whole of their lives in the water. Aquatic larvae have a tracheal system which does not differ much from the normal pattern, except that the stigmal openings are either closed off or are initially obliterated and later become permeable. An example of the first type is the Water Veneer moth, *Acentria nivea* (Pyralidae), which is found commonly throughout Europe, including Britain, and also occurs in North America. The adults lay their eggs on various water plants, and on hatching the larvae live under water, remaining submerged for several months in a water-filled case made from leaf fragments. The spiracles are apparently impermeable and the larva breathes through the skin. When fully grown it pupates in an underwater cocoon attached to its foodplant. A septum divides the cocoon into two compartments, one of which contains the pupa and also a large amount of expired air produced by it.

An example of the second type of behaviour is found in another pyralid, *Nymphula nymphaeata*, which is one of the China-Mark moths. The female

Above: The caterpillar of the Satin moth, Leucoma salicis *(Lymantriidae), which lives on willow and poplar leaves.*

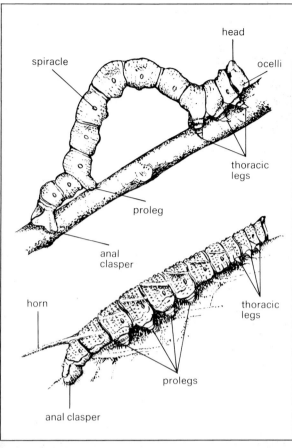

of this species lays her eggs in batches on the under-water surface of various waterplants, preferring the still water of ponds and lakes. When young the larva has impermeable stigmata, and breathes the oxygen dissolved in the water through its skin. Later the stigmata open and from then on it can breathe atmospheric air. It lives in a flat oval case made of pieces of leaf and attaches its case to the plant with silk. At times the case floats to the surface and is rocked from side to side by the larva, presumably to take in air.

A few related species have developed tracheal gills or blood-gills and are considered to be the most highly adapted aquatic Lepidoptera. Most aquatic Lepidoptera belong in the Nymphulinae, but isolated cases are also known in the Noctuidae and Arctiidae.

The abdomen of the caterpillars of many lycaenid butterflies have special myrmecophilous glands. There are generally three of these which are arranged in a triangle on the upper part of the abdomen, one being on the seventh and two on the eighth segments. The anterior gland produces a sweet liquid, while the posterior pair produce a scented secretion which is greatly relished by ants. Having located the larva the ants proceed to stimulate the anterior gland by caressing it with their antennae. This then swells up and secretes droplets of honeydew which are avidly sucked up by the ants. Thus, these lycaenid larvae get protection and live in a friendly symbiotic

Left : Diagram showing the parts of a caterpillar.
Above : The Looper or Inchworm caterpillar of a geometrid moth ;
Below : The caterpillar of the Lime Hawk-moth, Mimos tiliae *(Sphingidae).*
Below : Young larvae of the Privet Hawk-moth, Sphinx ligustri *(Sphingidae), feeding on a privet leaf. The mature caterpillars are quite different from these young ones both in shape and colour.*

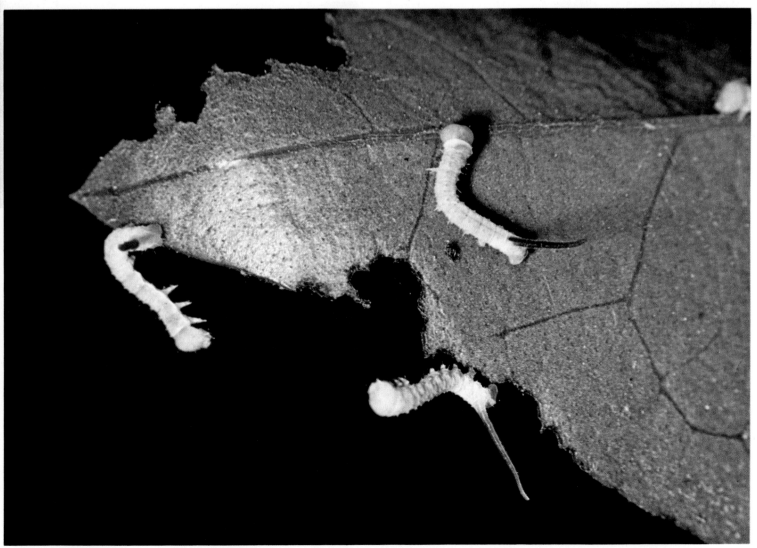

relationship with the ants, each drawing some benefit from the association. However, in some cases the benign relationship does not last but changes to a parasitic one. This is seen in the life-cycle of the Alcon Blue, *Maculinea alcon* (Lycaenidae), which is found from northern Spain and France across Europe to central Asia. These large and elegant butterflies lay their eggs on the flowers of the marsh gentian (*Gentiana pneumonanthe*). The young larvae live inside the flower, finding protection at night and during rain, when the flower closes up, and eating its internal organs—particularly the ovary. Several larvae may be found living side by side; the cannibalism that marks out their relative the Large Blue, *M. arion*, is not found in *M. alcon*. The caterpillars spend all summer and the early autumn inside the flower; and then they seem to disappear into thin air, while those kept in captivity in breeding cages all die at this point.

For a long time no explanation could be found for either phenomenon. Not until 1916 did the British naturalists Harold Powell, T. A. Chapman and F. W. Frohawk begin to unravel the mystery of these and other lycaenids by dint of long observation and ingenious breeding experiments. Powell, using information provided by Chapman for *M. arion*, began to offer the larvae of *M. alcon* that he had bred, chopped ant-larvae to eat, when they reached this critical stage in their development. He found that this was just the food they required, and larvae fed on it completed their development normally. It was then found that the ants take the larvae of *M. alcon* from the gentian flowers and carry them off to their nest, where they are reared together with the ants' own larvae. It is almost certain that the ants feed the lycaenid caterpillars with drops of a liquid that they regurgitate from their mouths; but it is also quite certain that the butterfly larvae suck out the contents of the bodies of the host larvae, leaving only the hard skins (unlike *M. arion*, which eats these too). The surprising discovery was then made that the caterpillars of *M. alcon* have only two moults, since none takes place in the ants' nest. As the caterpillar grows, its cuticle stretches, becomes thinner and more transparent, until it seems to be on the point of bursting. All this only happens when pupation is about to take place that is to say, in the ants' nest, at the end of spring. The pupal stage lasts about 20 days.

Ants that have been found in association with *M. alcon*, either in nature or in artificial nests, are *Myrmica scabrinodis*, *Tapinoma erraticum* and *Tetramorium caespitum*.

Not all lycaenids are obliged to live in ants' nests in order to survive. Some species have a life-cycle that does not depend on ants at all, while caterpillars of other species may be invariably accompanied by ants though the latter are not essential for their development.

The internal anatomy of the lepidopterous larva is relatively simple. The digestive system or alimentary canal consists of three primary divisions that have different structures and functions. The first section is known as the stomodaeum or fore-gut and connects the mouth with the middle segment of the intestine, the mesenteron or mid-gut. At the junction between the first and second parts of the alimentary canal lies the cardiac valve, equipped with numerous muscle fibres which help it to close. Another valve, called the

pyloric valve, divides the mesenteron from the proctodaeum or hind-gut, which is the final section of the alimentary canal and terminates at the anus.

The stomodaeum and proctodaeum are both derived from ectodermal structures and have a luminal surface covered with cuticle which is continuous with the cuticle covering the whole outer surface of the caterpillar's body. The mesenteron, however, is derived from endoderm, and therefore has no cuticular lining. Almost all the processes of digestion and absorption take place in the mesenteron.

There are a large number of glands connected with the alimentary canal; most of them have a digestive function. By far the most important complex of glands, however, is the salivary complex. The salivary glands are divided according to their location into mandibular and labial glands; in the Lepidoptera they are only present in the larval stage.

The two labial glands or silk glands are situated in the head close to the labium or lower lip and open on the spinneret. Sometimes they are so well developed as to extend throughout the whole length of the caterpillar. They are sometimes complexly folded and, in the Silkworm, *Bombyx mori*, and some of the saturniid silk moths, their length if stretched out would be several times that of the body. In a mature Silkworm they can account for more than 25 per cent of the body weight. Silk plays an important role in the daily life of nearly all caterpillars, except in some of the leaf-mining species, and is secreted almost continuously. It has many uses: it can provide a foothold on slippery leaves, it acts as a life-line at times of emergency, it gives the caterpillar protection in the form of silken cases and tubes, or it can be used for spinning leaves together to make shelters, and when the larva is fully grown it may be required for constructing a cocoon. In section, the silk gland is seen to be divided into three parts, each one having a different function. The distal part furthest from the duct opening is thin; this is the part in which silk is produced. The second part, much wider in diameter, also secretes silk and is known as the reservoir. The third part, nearest the mouth, is the duct of the gland; it is narrow and tortuous. The two ducts from the silk glands meet in a single common duct; one part of this is modified and supplied with large muscle-bundles which regulate the silk flow. The whole apparatus terminates in the spinneret on the labium.

Silk is produced in liquid form, and solidifies on contact with air by a process of oxidation and not by a simple drying process as was previously thought. Indeed, silk will solidify even if it is secreted into water. It varies in colour and texture with different species and is often mixed with frass or some other extraneous material in construction of the cocoon. The silk of the Cultivated Silkworm is typically a lustrous pale yellow, but can vary from white to yellow, the quality and colour depending very much on the living conditions and food of the larvae. The best silk for commercial purposes comes from larvae fed on fresh mulberry leaves, which supply amino-acids important for its formation and carotinoid pigments for its coloration. In an emergency the larvae can also be fed on garden lettuce, endive and strawberry leaves, but the silk is then inferior. Except for some loose surface strands, the cocoon of the Silkworm is constructed from a continuous thread. When the cocoon is finished the larva is completely

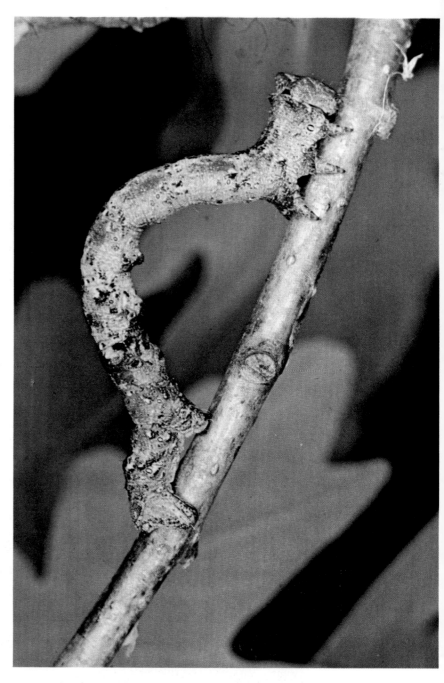

sealed in with no access to the outside world, somewhat like the embryo of an egg. It is thus secure from ants and similar prying predators. It soon changes to a pupa and in about two weeks the adult develops. The wall of the cocoon is much too dense and the silk too strong for the imago to force a way through, and to get free it secretes a solvent which weakens the silk, usually at one end of the cocoon, and then squeezes out.

Saturniid moths which are reared for their silk in India include *Antheraea pernyi* and *A. roylei*, which produce tasar silk, and the Eri silk moth, *Samia cynthia*. These moths are much larger than the Cultivated Silk moth and their larvae make bigger cocoons, but the silk is generally coarser and the length of the thread shorter. The larvae of Tasar silk moths are mostly polyphagous, but oak (*Quercus*), is one of the chief food plants. Experiments have been carried out with hybridization of races of these moths from different parts of India and China, and improved strains have been bred. The natural shades of tasar silk are light, medium and dark brown, but the

Above: The typical appearance of a geometrid caterpillar in motion along a twig.

colour range now extends to silvery white, grey and yellow. The thickness and evenness of the silk is of primary importance and it has been found that larvae fed on the foliage of *Shorea robusta,* a timber tree, produce a superior silk.

Silk is produced as a fine, tough thread, a cross-section of which shows an inner fibre consisting of fibroin and an outer sheath of sericin. Both fibroin and sericin are proteins. The inner fibre of the silk thread is actually double, and consists of the two threads secreted by the paired silk glands compressed into a single thread as they pass through the common duct. The walls of the common duct regulate the thickness of the thread and the quantity of silk secreted, and they can even cut off the thread when necessary.

Lepidopterous larvae are mainly phytophagous (plant-eaters), and attack nearly all plants from microscopic algae and fungi to tall trees. Plants containing substances poisonous to man or animals may be coolly devoured by the caterpillar without the slightest ill-effect. Every part of the plant, from the

roots to the terminal shoot, flowers and seeds, is used for food; nor is the fruit spared whether it is sour or ripe. It is true to say that while adult butterflies and moths are only very rarely harmful, their larvae are among the most voracious of insect pests. Some of the larger caterpillars devour a leaf, beginning at the edge of it and often advancing so fast that within a few minutes all that is left is the stem and the thickest veins.

Heavy infestation by leaf-eating caterpillars that are nearly fully grown can lead to rapid defoliation on a scale exceeded only by a swarm of locusts. Some of the tunnelling wood-eating species, such as the Goat moth, *Cossus cossus* (Cossidae), may spend more than two years in the larval stage. During this time they make great burrows and ingest a volume of woody food that is enormous in comparison to the limited food value they can obtain from it. This is presumably because of the slow process of converting cellulose into useful body-building carbohydrates.

Among the phytophagous or plant-feeding larvae there are some which are monophagous and feed

Above: The ornamentation of lepidopterous larvae is sometimes extremely rich. In this silkmoth larva, both the coloration and the magnificent spiny appendages are eye-catching.

only on one species of plant, others are oligophagous and will feed on the plants of a particular genus or family, and others are polyphagous and will feed on a wide variety of plants. The most important pests generally belong to the latter two groups.

While most lepidopterous larvae feed on the tissues of living plants, many have found other sources of food supply. Some attack substances of plant origin, such as dead wood, cork, dried fruit, herbs and fungi, stored flour and other cereals and also chocolate. Numerous species, especially among the Microlepidoptera, are saprophagous, living on dead animals or material of animal origin and decaying vegetable matter. This particularly applies to the clothes moths of the family Tineidae, whose depredations involve insect collections, stuffed animals, hides, hair, horn, feathers, fur and wool. Most people

are anxious about the damage that might be caused by clothes moths that get into houses from time to time. However, in recent years this risk has greatly diminished with the increased use of synthetic fibres on which they cannot thrive.

A few rare instances are known of species which have become coprophagous and live on the excrement of other animals. This habit has led to the development of a remarkable relationship between some pyralid moths and the sloths of South America. One of these species is the Panamanian Sloth moth, *Cryptoses choloepi*, which is associated with the three-toed sloth (*Bradypus infuscatus*). The adult moths live a quiet existence in the fur of the sloth, mating and perhaps drinking the secretions from sebaceous glands among the sloth's hairs, while patiently waiting for it to descend to the forest floor to defecate.

Above : Some geometrid caterpillars resemble slender, dry twigs, and can stay motionless for long periods to enhance the effect.
Right : The larva of the Lime Hawk-moth, Mimas tiliae, *(Sphingidae), attacking a leaf. Note the extensive inroad already made into the leaf. This hawk-moth is a common moth in Europe and is often found in towns and cities.*

This the sloth does about once a week, scraping a shallow depression in the ground in which to deposit its pellets and then covering them with leaf litter. The waiting female moth promptly takes a short flight and lays her eggs on the dung, and then returns to her hideaway in the sloth's fur.

A number of species of galleriine moths of the family Pyralidae live on wax forming the honeycomb in the nests of bees. The best known of these is the Wax moth, *Galleria mellonella*, whose larvae can be a nuisance when they get into domestic beehives.

Predation is relatively rare among the Lepidoptera, and occurs only in isolated cases. The larvae of some acontiine (Noctuidae) and phycitine (Pyralidae) moths are specialized feeders on scale insects (Coccoidea) and plant hoppers (Homoptera), and a geometrid pug moth of the genus *Eupithecia* lives on fruit flies (*Drosophila*) in Hawaii. The immature stages of ants are preyed upon by lycaenid butterflies in many countries. Once the lycaenid caterpillar is in their nest the ant occupants seem oblivious or unable to take effective action to prevent the intruder devouring some of their brood.

Aphids or greenfly are often inadvertently eaten by plant-feeding lepidopterous larvae and are sometimes taken deliberately. Cannibalism occasionally occurs, and perhaps the most notorious cannibal is the Dunbar moth, *Cosmia trapezina* (Noctuidae), which is normally a leaf-eater but will readily attack and devour other larvae of either its own or other species which it may come across.

Whatever their food, caterpillars are tireless eaters. Sometimes they are compelled to make long and difficult journeys in search of fresh supplies. For better or worse, nature has provided them with all the necessary equipment for covering considerable distances and for holding on to almost any support. The prolegs are vital in this context, and the sucker-like anal claspers can keep the hind end of the caterpillar anchored to its support; but for this, the back end would trail along the ground and would certainly prevent the caterpillar from calmly feeding head downwards. This system of attachment and adhesion is supplemented by the silk thread secreted by the caterpillar through its spinneret. Also, when some caterpillars are disturbed from their feeding place and fall, they do not plummet to the ground but make a controlled descent by spewing out a silk life-line through the spinneret. When the danger has past, they haul themselves up this life-line by means of their mandibles and thoracic legs and return to the feeding place.

The prolegs are operated by special muscles that have a purely locomotory function: the longitudinal muscles of a particular segment, in the dorsal position, contract simultaneously with the vertical muscles of the preceding segment (which raise the proleg from the ground) and with the ventral longitudinal muscles of the following segment. This gives rise to a contraction-wave, which allows each pair of legs to move forwards a short space. When the caterpillar needs to take a firmer hold during its advance, it relaxes the vertical muscles so that the proleg becomes turgid from the internal pressure of haemolymph. The sucker-shaped tip can then take a firmer hold on the support.

Some caterpillars are very slow movers, while others move fast and sometimes in jerks. The characteristic pattern of movement for a particular species depends largely on the number and position of the prolegs. Among the most typical modes of progression, one must mention that of the geometrid caterpillars, also known as loopers or inch-worms—names derived from the way in which they move along a surface as if measuring it inch by inch. These caterpillars lack one or more pairs of prolegs on the foremost abdominal segments. To move along, they fix their anal prolegs or claspers to the surface and push the body forwards. They then hold on to the ground they have gained with their thoracic legs, and pull their abdomen forward by bending it upwards like a hoop. Then the front of the body moves forward again, and the hind end is brought up to it again.

Another characteristic way of moving is found in those caterpillars that build cases out of various types of material, in which they spend the whole of their

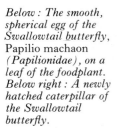

Below : The smooth, spherical egg of the Swallowtail butterfly, Papilio machaon *(Papilionidae), on a leaf of the foodplant. Below right : A newly hatched caterpillar of the Swallowtail butterfly.*

Right : A caterpillar of the Scarce Swallowtail butterfly, Iphiclides podalirius *(Papilionidae), preparing for pupation. Note the supporting silken girdle encircling the upper part of the body.*

larval stage and in which they often pupate. In order to move, these caterpillars only use their thoracic legs, stretching their body out to a great length and protruding the front of it out of the case. They fix themselves in position with their legs, and then pull the rest of their body forwards, taking the case with it. The case is held firmly in position over the hind end of the body by the hooks on the abdominal prolegs.

Tireless in their work, caterpillars nonetheless sometimes take a well-earned rest. Some will lie completely motionless, merging perfectly with their surroundings; others attach themselves in place with a silk thread and hold on with their prolegs for additional safety.

The caterpillar represents the growth stage in the metamorphosis of a lepidopteran, and body growth is one of the most important functions of this post-embryonic stage of a butterfly or moth. The constraints imposed by the exoskeleton mean that this growth must be an interrupted process. The caterpillar is entirely covered by a more or less rigid protective coat consisting of its cuticle or integument, which is the outer part of the skin and is thin and flexible between the segments to allow movement. Any growth in length or volume must therefore involve the casting off of the old covering and its replacement by a larger one. This is the phenomenon of the moult and is termed ecdysis. It seems a simple process but in fact in many ways it is very complex. The period between moults, during which the larva is feeding and filling out its new skin, is called an instar. Most lepidopterous larvae have four or five instars, but there is considerable variation depending on the species and prevailing conditions.

When the caterpillar is obliged to cast off its old cuticle, which by now is preventing any further growth, a special secretion is produced by the so-called exuvial glands and by the hypodermic cells; this secretion accumulates in a very thin layer between the cuticle and the hypodermis which originally produced the cuticle. The secretion contains enzymes which partially digest the old cuticle, so that the covering that is cast off is thinner than the new cuticle that now grows on the caterpillar's body, produced by the hypodermis and laid down between the hypodermis and the old cuticle.

At this point, then, the caterpillar is clothed in a new exoskeleton, while its old cuticle is still in place, though separated from its body by a thin layer of exuvial fluid. It is now necessary to get rid of the unwanted old covering. The caterpillar dilates its body wall, using muscular contractions to increase the pressure of its internal fluids; it accumulates its intestinal contents at particular points in its alimentary canal; and it introduces large amounts of air not only into its digestive system but also into its complex and highly developed tracheal system. The combined effect of some or all of these manoeuvres results in a considerable dilatation of the caterpillar's body, and thus the exertion of considerable pressure on the old cuticle. The cuticle tears along the lines of least resistance, enabling the caterpillar to escape from the structure which now becomes known as the exuviae. The now useless discarded cuticle or exuviae may be abandoned altogether, but more often it is eaten by the caterpillar.

The mechanisms involved in the moult do not seem to be of any great complexity; but in fact they are. One feature that is very complicated indeed, and in fact is not yet properly understood, is the question of what actually triggers the larva off to start the process of moulting.

In Lepidoptera, as in all exopterygote insects, the moult is governed by a well-regulated pattern of activity of secretory areas directly controlled by the central nervous system, particularly the brain. This latter, which lies in the cranial capsule and above the pharynx, is divided into three parts: the protocerebrum, the deutocerebrum and the tritocerebrum. The protocerebrum is the largest component of the brain, while the tritocerebrum is the smallest.

The protocerebrum, and particularly the so-called *pars intercerebralis*, are of particular importance in the moulting process. In these areas there is a concentration of nerve-cells producing a special neural secretion; they show cyclic activity, and respond to various stimuli. The hormone produced by these neuro-secretory cells travels through nerve-channels to two bodies known as the *corpora cardiaca*, lying behind the brain and connected to it through nerves. The *corpora cardiaca* thus act as centres for receiving and concentrating the neural secretion produced mainly by the *pars intercerebralis* of the brain. However, apart from their function as centres for the accumulation of secretions produced elsewhere, they also produce their own secretions.

The hormones produced or concentrated by the *corpora cardiaca* enter the caterpillar's circulatory system, and stimulate the prothoracic glands. Despite their name, these glands may be located in various parts of the body; they secrete a non-specific hormone known as the moulting hormone or ecdysone.

Another pair of structures known as the *corpora allata* are connected to the *corpora cardiaca*, and occasionally fused more or less completely with the latter. The *corpora allata* also secrete a hormone, known as the juvenile hormone.

While the *corpora allata* are producing juvenile hormone and secreting it into the circulation, ecdysone induces pre-imaginal growth and moulting. In the absence of juvenile hormone, however,

Left: The tough, silken cocoon of the Oak Eggar moth, Lasiocampa quercus *(Lasiocampidae). In very cold climates the pupa may spend two whole years inside this refuge before the moth emerges.*
Right: The pupa of the Camberwell Beauty butterfly, Nymphalis antiopa *(Nymphalidae). The bristly remains of the caterpillar's shed skin can be seen on the branch.*

ecdysone sets in motion the process of metamorphosis, in which the caterpillar is transformed into a fully-formed adult or imago.

The production of juvenile hormone, which slows down or ceases altogether at the end of the caterpillar's period of growth, resumes once again in the imaginal stage. In the adult butterfly or moth, juvenile hormone acts to stimulate the development of the gonads, in which germinal cells are made.

The number of moults varies between different Lepidopterous species and families. There may even be variation between members of a single species; the female often moults once or twice more than the male. Excessive numbers of moults may take place for various reasons: there may be genetic factors, a shortage of food or temperature variations.

Every caterpillar, after completing the appropriate number of moults, stops feeding and prepares to undergo the succession of far-reaching changes that will eventually culminate in its transformation into an imago. Among Lepidoptera, there is an enormous difference between the morphology of the caterpillar and that of the adult. The definitive organization of the adult's body is achieved during an intermediate phase between the caterpillar and the imago: this stage is known as the pupa or chrysalis.

The pupal stage is a dangerous period for the insect as it is wide open to attack by all sorts of enemies, with no possibility of aggressive defence or flight. Therefore it is only natural that when the time comes to pupate, many caterpillars seek out some form of protection, such as a hiding-place in the

crevices between stones or under the bark of a tree, under moss, in the soil, or among dense vegetation. Some start by building themselves a secondary refuge or cocoon before pupation. The cocoon varies in shape, thickness, size and colour between one species and another, and is made from silk and other secretions mainly produced by the labial glands. In addition to silk, the cocoon may consist of all sorts of extraneous materials.

Not all Lepidoptera build themselves a cocoon when they reach the end of their larval life. Many pupae remain without any form of defence other than their cryptic appearance and coloration. These naked pupae may either hang from the substratum, or be attached to it by a kind of girdle. In the former case, the pupa hangs head downwards from a pad of silk which is fixed to the substratum; the caterpillar first takes up its position here holding on with its anal prolegs, and as the skin splits, the tail of the pupa becomes hooked in the same position. In the latter case, the caterpillar fixes itself to the substratum with a tough silk thread round its own body, like a safety-belt, usually around the insect's thorax but occasionally round the head. The chrysalis, once it has shed the larval skin, remains attached to the substratum by the same silken girdle; it also attaches itself by the terminal part of the abdomen.

During the pupal stage the insect remains virtually motionless; some make no movement at all, while others are capable of a small degree of movement if disturbed, or when metamorphosis into the adult form is near.

The duration of the pupal stage varies greatly between species, and even within a species its length can vary according to prevailing conditions. In some Rhopalocera and Microlepidoptera it may last less than two weeks if conditions are optimal. In many species living in the temperate regions, the pupa overwinters and the adult does not emerge until the spring. Some species, including many cossids, may spend two or more years as pupae, especially when living in dry and arid regions. Pupae of some of the hawk-moths (Sphingidae) and other species which pupate in the soil may be buried 15 centimetres (6 in),

or more below the surface and are virtually entombed during periods of drought, when the ground becomes rock-hard. They can usually detect when conditions are hopeless for the emergence of the imago; for example, if the season for egg-laying has passed and there is no food available for the larvae. In this case they will remain quiescent until the following season, emerging when conditions are more propitious. Instances of this delayed emergence are most frequent in the larger species of Heterocera although, remarkably, cases are known of even very small Micro-moths which are more susceptible to desiccation, emerging two or more years after pupation.

In many respects the pupa bears a mummified resemblance to the adult, and might be regarded as a highly specialized first instar of the adult stage. In it the three regions of the body—head, thorax and abdomen—can easily be made out, and the form of the compound eyes, antennae, proboscis (when present), legs, wings and the segments of the abdomen are discernible in shallow relief. The cuticle is generally tough and hard, and often bears protuberances and spines, but is usually devoid of hair.

On the head, the compound eyes are well developed and can be easily seen, though they are almost certainly without function at this stage. With few exceptions, the antennae are bent back along the ventral surface, between the legs and the wings and are stuck together, usually along the body mid-ventrally.

When present, the proboscis is bent back mid-ventrally and attached to the body along its length, except for some sphingid moths in which it is curved in shape and free. The mouth is without function, except in a few species that swallow large volumes of air into their digestive tract as a preliminary to hatching out into the imaginal stage. Only in the primitive micropterigoids and eriocranoids are the mandibles well developed and even functional.

The thorax of the pupa shows all the structures that will eventually be present in the adult. The legs appear to be squashed together, with the tarsi of the last pair sticking out between the first two pairs.

These and the following two pages depict the life-cycle of the Two-tailed Pasha butterfly, Charaxes jasius (Nymphalidae). Above left: The egg of the Two-tailed Pasha butterfly. Above: The newly hatched caterpillar. Right: The almost fully grown caterpillar. Overleaf: The pupa, or chrysalis, which eventually splits open to allow the adult or imago to emerge and expand its wings. Overleaf right: The adult which is newly emerged.

The rudiments of the front pair of wings, arising from the mesothorax, are enclosed within triangular coverings. These are much shorter and narrower than the adult's wings, since in the pupal stage the wings are folded back on themselves several times over. The second pair of wings, arising from the metathorax, is hidden by the first pair.

The abdomen is formed of 10 segments or somites; they do not bear prolegs like the abdominal segments of the caterpillar, and often not every segment can be made out. The stigmata or openings of the respiratory tract on the body surface are clearly visible; they are equal in number to those of the caterpillar, though the last pair is rudimentary and functionless.

The anal orifice is closed, and marked only by a depression on the surface of the body. The same applies to the zone corresponding to the male and female genital openings. In the male, there is a single depression, while the female may have one or two depending on the species. The last abdominal segment often carries a cremaster; this is an arrangement of hooks used for attaching the pupa in position.

The more primitive Lepidoptera, for example the Micropterigidae and Eriocraniidae, generally have a free or exarate pupa, which has all its abdominal segments, except the last three, free and more or less mobile, as are the wings, the legs and the other body appendages. Other species have a so-called incomplete pupa, with wings and appendages partially free but immovable. The pupa is nevertheless able to move a certain amount by vigorously contorting its abdomen, and in many species has backwardly directed spines on the surface of its body which hook in the surface and aid it in wriggling towards the exit hole. In this type the pupa partially extrudes from its shelter during emergence of the adult. This is a characteristic of most of the Microlepidoptera.

A third type is the so-called obtect pupa, characteristic of the more highly evolved Lepidoptera, including the Rhopalocera and most other Macro-lepidoptera. Here, the wings and other appendages are glued together and to the body itself, and most of the abdominal segments are usually rigidly fused together. This means that the pupa is almost completely unable to move—all it can do is to make a small movement with part of the abdomen. These pupae usually do not have dorsal spines but the cremaster is well developed.

As we have said, pupae take differing periods of time to reach maturity depending on their species or on environmental factors. In chrysalids with an almost transparent cuticle, it is possible to make out the internal structure of the adult insect a few days before the adult emerges; even the colours of the wings that still lie folded around the body can be distinguished. As a rule, however, the cuticle becomes progressively darker as the pupa matures.

Eclosion, or emergence of the adult insect, is a process that may be simple or less so depending on the type of pupa and on the structures that surround it. The simplest case is that of the naked pupa living in contact with its environment. Here, the adult insect, lying within the pupal skin, merely has to rupture the cuticle and protrude its head. This requires a considerable effort, however; it may be made easier by the presence of tough pointed spines on the head, which act as an awl and bore a way through the cuticle.

Matters are less simple if the pupa is underground or within the tissues of a plant, particularly if it is also enclosed in a cocoon. Some subterranean pupae are free or incomplete and when eclosion is near, they work their way to the surface; whereas in the case of obtect pupae the adult emerges below ground and has to struggle through the soil to the surface before it can expand its wings. Pupae living within plant tissues also generally emerge into the open at this stage; in most plant-mining species, of which there are a large number, no very great effort is required to bore through the plant tissues and come out into the open.

Below left : A cross-section of a cocoon, showing the pupa within it. Below : The pupa of the Convolvulus Hawk moth, Agrius convolvuli (Sphingidae) ; note particularly the long trunk-like proboscis lying free in its special case. This is a wide-ranging migratory species and the imago is strongly attracted to the scented flowers of white tobacco.

In those species that construct a cocoon, the silk thread of which it is made may be parted by pressure exerted by the insect's head, or cut with special structures evolved for the purpose, or partially digested by the secretions of special glands.

There is a great deal of variation in the methods used for splitting the pupal covering. The cuticle of the obtect pupa usually remains intact except for a split along the medial fracture line. The cuticle of an incomplete pupa, on the other hand, splits along numerous fracture lines demarcating the component parts of the thoracic appendages before the adult can escape. Once it has released its head from its coverings, by whatever means, the adult butterfly or moth swallows a large quantity of air through its mouth; then it uses its legs and the contortions of the rest of its body to complete the process of eclosion in a relatively short space of time. After this, numb with exhaustion, it remains firmly attached in position close to the pupal exuviae. The wings are still mere shapeless, limp sacs, giving no hint of their eventual often splendid pattern; so that the adult butterfly or moth at this stage is a pitiful-looking creature. Blood is pumped into the wing veins and soon the wings expand and dry out; the adult now takes on its definitive appearance. Before flying off the insect gets rid of the waste products that have accumulated during its pupal life. In the chrysalis, the anal aperture is still closed; as the products of excretion are generally conveyed, in all insects, to the last part of the digestive tract, this region accumulates a large amount of material that can only be got rid of once the anal aperture is functioning. The waste material, which contains a large amount of water, is known as meconium; it is occasionally bright red in colour, and its excretion by a large number of newly-emerged individuals almost simultaneously has on occasion given rise to legends about a 'rain of blood'.

After expelling its meconium, the butterfly or moth can take flight. Looking at an adult insect of this kind, one can appreciate the enormous changes that have taken place during metamorphosis. The imago bears not the slightest resemblance to the caterpillar: the mouth-parts have changed from the chewing to the sucking type: the wings have appeared: every part of the insect's structure has altered. All these phenomena are not peculiar to Lepidoptera but are typical of holometabolous insects. In hemimetabolous insects, on the other hand, most larvae resemble the adult, and the transformation from larva to imago takes place in a series of gradual stages rather than all at once.

Above: The short-bodied pupa of the Scarce Swallowtail, Iphiclides podalirius *(Papilionidae). The two pointed projections at the head end, and the tough strand of silk supporting the pupa are of particular interest.*

Mimicry and
Other Protective Devices

Lepidoptera are food for a multitude of predatory animals including: mice, moles, bats, birds, ants, beetles, ichneumon, braconid and chalcid wasps, robber-flies (Asilidae), dragonflies (Odonata), mantids, scorpions and spiders. There are also a few aquatic moths which have yet another set of enemies. This army of adversaries has triggered the evolution of many different protective devices by Lepidoptera.

One of the most fascinating strategies, from an evolutionary point of view, is the adoption of a mimetic life-style. Mimicry involves a deceptive or mimetic resemblance of one organism to another or to some inanimate object to such a degree that enemies are deceived by the mimic. The similarities between a mimic and its model are usually of a visual type but often include behavioural characteristics and may also involve sound and smell. No one has ever questioned the seemingly close resemblance between many Looper caterpillars (Geometridae) and twigs, between the under surface of the Indian Leaf butterfly, *Kallima inachus* (Nymphalidae), and a leaf, or between females of the African Mocker butterfly, *Papilio dardanus* (Papilionidae), and the various other species of unpalatable butterflies which they mimic. What has been disputed is the proposition that predatory enemies, particularly birds, are also deceived by these similarities and that predation is the selective pressure which has brought about the evolution of mimetic resemblances. Indeed, there has been controversy ever since the end of the last century when Bates and Müller first put forward their ideas about mimicry.

Experiments have demonstrated that insectivorous birds are able to distinguish between different shapes, colour-patterns and colours (at least in the yellow to red part of the spectrum). Many birds reject as food unpalatable insects which they quickly learn to associate with a particular warning colour-pattern but palatable mimics of unpalatable model species often deceive these birds. Fairly recent chemical

analysis has shown why some predators find certain insects unpleasant. Hydrogen cyanide, and various histamines, acids, alkaloids, and heart poisons (cardiac glycosides) have been extracted from the tissues of arctiid, hypsid, ctenuchid and zygaenid moths as well as from ithomiine, heliconiine, danaine, acraeine and some papilionid butterflies. The African Swallowtail butterfly *Papilio antimachus*, can contain enough heart poisons to kill several cats. Poisons have been found concentrated in the wings of the Monarch butterfly, *Danaus plexippus* (Danaidae), a device which tends to ensure that an investigating bird will receive the strongest chemical rebuff after pecking the least vital part of the butterfly.

Most unpalatable Lepidoptera extract their poisons directly from plants which they eat. Caterpillars of the heliconiine Passion-flower butterflies feed on the foliage of Passion-flowers (Passifloraceae), a family of plants normally poisonous to birds and other vertebrates; some American Tiger-moths (Arctiidae), feed on milkweeds (Asclepiadaceae), a family poisonous to cattle, while many South American ithomiine butterflies feed on Solanaceae, a family which includes deadly nightshade (*Atropos belladonna*), among its species. The North American and temperate Old World Garden Tiger moth, *Arctia caja* (Arctiidae), extracts pyrrolizidine alkaloids from *Senecio* plants, and is consequently highly distasteful to birds. Plant poisons may be stored unmodified by the caterpillar, to be passed on to the adult insect, or modified to form another equally unpleasant chemical.

A few toxic Lepidoptera seem to be able to manufacture their own poisons. For example, zygaenid moths store a chemical similar to hydrogen cyanide in their tissues and are not dependent on a particular poisonous group of plants.

Many species have evolved unpleasant advance-warning smells which ward off predators and act as a survival mechanism. The Crimson Speckled moth, *Utetheisa pulchella* (Arctiidae), of the Old World and the Bella moth, *U. ornatrix*, of America produce a pungent froth from thoracic glands just behind the head if provoked. Many other species have similar

Left: The Lappet moth, Gastropacha quercifolia *(Lasiocampidae), arranges its wings so as to resemble a bunch of dry leaves when it is at rest.*

Left : When at rest
with its wings closed
vertically, the Peacock
butterfly, Inachis io
(Nymphalidae), blends
with the background.

glands, and often have a visual reminder to predators in the form of a red, yellow or orange patch of scales marking the position of these glands on the thorax.

Batesian mimicry, named after the English evolutionist H. W. Bates, involves an unpalatable model which is mimicked by one or more relatively palatable species. Bates, who drew his conclusions from studies made along the banks of the Amazon, suggested that in a group of closely similar, boldly patterned species flying together one of them is likely to be unpalatable, and that the remaining species will be palatable mimics. Though much of Bates's theory is substantiated, he was only partly right because a number of his palatable mimics are now known to be unpalatable.

Some of the most convincing laboratory proof of Batesian mimicry has been produced in the United States by J. V. Z. Brower and L. P. Brower. In one series of experiments, inexperienced Florida Scrub jays were offered unpalatable Monarch butterflies, *Danaus plexippus*, a species which eats plants containing heart-poisons. The birds ate the Monarchs, then vomited and soon learnt to avoid further Monarchs. The jays were then offered specimens of the supposed mimic of the Monarch, the relatively palatable Viceroy butterfly, *Limenitis archippus* (Nymphalidae), but resolutely refused to eat them. Jays in a parallel experiment which had no experience of Monarchs willingly ate most Viceroys offered to them. These experiments demonstrated convin-

cingly the protective resemblance of the palatable Viceroys to the unpalatable Monarchs. It was found later that some milkweeds contain no heart-poisons and that inexperienced American Blue jays would happily eat Monarchs whose caterpillars had fed on non-poisonous milkweeds, demonstrating that it was the heart-poisons which the jays had learnt to avoid.

The Monarch butterfly is typical of many warningly coloured, day-flying models in that it reinforces its colour-pattern with a deliberate flight, a strategy copied by its mimics. Also typical is the tough leathery nature of the body which can withstand pecking by a bird without fatal results to the butterfly. Arctiid and zygaenid moths are similarly resilient insects.

Communal roosting is a fairly common feature of noxious butterflies, by resting together they concentrate and emphasize their warning colours and smells. The Monarch is probably best known in this respect; it overwinters in large roosts in California and Florida and in a vast roost in Mexico which has been recently discovered. Polymorphism (the presence of two or more genetically controlled colour-forms in one species) in Passion-flower butterflies has resulted from predation pressures which have induced the evolution of changes in colour-pattern in these species towards those of even more unpalatable ithomiine butterflies.

Most Clearwing moths, (Sesiidae), mimic venomous wasps and bees, and many even buzz like bees.

*Right : The Dark
Sword Grass moth,*
Agrotis ipsilon
*(Noctuidae), when
resting holds its wings
horizontally and partly
overlapping each other
becoming almost
invisible among lichen
on a tree-trunk.
Overleaf left : A
specimen of an
Oriental Leaf butterfly*
Kallima inachis
*(Nymphalidae),
displayed with
outspread wings to
show the brilliant
colours of the pattern
adorning the upper
surface of its wings.
Overleaf right :
Another specimen of
the same species with
its wings closed
together. In this
position it bears a
remarkable resemblance
to a dry leaf.*

Above : Two Red-underwing moths of the genus Catocala *(Noctuidae). The coloration and pattern of the forewing makes them almost invisible against the back-ground of a tree-trunk.*

The Costa Rican ctenuchid moth, *Amycles anthracina*, bears an astonishing resemblance to the wasp *Chartergus apicalis*. Sound is used by the palatable North American Banded woolybear moth, *Pyrrarctia isabella* (Arctiidae), to deceive bats; it produces pulses of sound at a pitch identical to the warning sounds emitted by many unpalatable nocturnal Tiger-moths. *P. isabella* is consequently a sound mimic of the distasteful Tiger-moths. The buzzing sounds made in flight by the European Lunar Hornet Clearwing, *Sphecia bembeciformis* (Sesiidae), and other Clearwings is also probably Batesian in function and will reinforce the already deceptive wasp-like appearance of these moths.

A particularly impressive Batesian mimic is the Swallowtail, *Papilio laglaizei*, whose model in New Guinea is the day-flying uraniid moth, *Alcidis agathyrsus*, relatives of which are known to feed as larvae on poisonous spurges. The only notable example of Batesian mimicry in Australian butter-flies is that between the Plain Tiger or African Monarch, *Danaus chrysippus* and its mimic the Six Continent butterfly, *Hypolimnas misippus* (Nymph-alidae).

Mimicking some behavioural trait of the model may be sometimes more effective than mimicking its appearance. For example, many Emperor moths (Saturniidae), some Tiger-moths (Arctiidae), and Birdwing butterflies, *Ornithoptera* (Papilionidae), will react like many bee species, raising their wings if provoked and curl the abdomen downwards displaying bands of black, yellow, orange or red.

It may not be necessary for a mimic to simulate the whole of a poisonous or venomous species. The wing-tips of the tropical Asian Atlas moth, *Attacus atlas* (Saturniidae), resemble a snake's head, as do many Hawk-moth caterpillars whose front end has a false eye on each side of the thorax. The Hawk-moth, *Panacra mydon* (Sphingidae), for example, mimics the head of the snake *Dendrophis picta*.

Müllerian mimicry, named after the German Fritz Müller, involves two or more unpalatable species, each of which advertises its noxious qualities with the same warning colours, smells, sounds or beha-viour. Having learnt to associate the warning features of one species with unpalatability, a predator will subsequently reject as food any other species having similar warning features.

Best known of the many examples of Müllerian mimicry are the several mimetic partnerships of unpalatable species of ithomiine, heliconiine and danaine butterflies and of pericopine moths in the South American tropics. The association of *Thyridia confusa* (Ithomiinae), an *Ituna* species (Danainae) and a *Notophyson* species (Pericopinae) provides one of the most impressive examples. Each of these species has basically transparent wings with almost identical greenish-black markings; the transparency being produced by reduction of the scales to hairs (*Thyridia*), reduction in the number of scales (*Ituna*) or by transparency of the scales themselves (*Noto-physon*). Within a single genus, the series of geo-graphically separated pairs of colour-forms of the Passion-flower butterflies *Heliconius erato* and *H.*

Lepidoptera and involves the deception of predators by an organism's resemblance to inedible background surfaces like tree bark, leaves, lichen, twigs and bird-droppings. There is not a great deal of experimental evidence in support of cryptic mimicry, but there is conversely hardly any evidence to show that predators are not deceived by the many cryptic devices which are so effective to the human eye. The forewings of thousands of moth species closely resemble tree bark on which the moths rest during the day. There are many examples in the nearly cosmopolitan noctuid genus *Catocala*, and at least one of these species, the White Underwing of North America, *C. relicta*, actively seeks out tree bark that matches its forewings in colour. Lichen-covered bark is mimicked by many noctuid and geometrid adults and by most lichen-feeding caterpillars. The European and Asian Lappet moth, *Gastropacha quercifolia* (Lasiocampidae), exposes both pairs of wings when at rest and bears a remarkable resemblance to a bundle of dead leaves.

Experimental evidence supporting the effectiveness of cryptic mimicry was produced by Kettlewell in his classical experiments with the genetically controlled dark and light forms of the Old World Peppered moth, *Biston betularia* (Geometridae). At rest, these moths remain motionless, with their wings partly outspread on the trunks or branches of trees. In non-industrial areas tree bark is likely to be partly covered with whitish-grey lichen on which the light speckled form of the Peppered moth will be well camouflaged, whereas in industrial areas the dark, nearly black form will blend more effectively into a background of dark soot-covered, lichen-free bark. Kettlewell's tests showed that bird predators do discover and eat a greater proportion of poorly camouflaged moths, whether they were pale moths on a dark background or dark moths on a pale background, and that selection by bird predation is very probably the cause of cryptic coloration in this species.

Many Swallowtail caterpillars (Papilionidae), are apparently bird-dropping mimics when young but become bark-like or leaf-like when they have grown beyond bird-dropping size. The caterpillars of the temperate Old World Alder moth, *Acronicta alni* (Noctuidae), is a bird-dropping mimic in its first two stages but later becomes a conspicuous black-and yellow-banded creature. Some small black and white adult moths like the European Chinese Character, *Cilix glaucata* (Drepanidae), are also thought to mimic bird-droppings.

Transparency is probably one of the most effective methods of merging into the background. Species with transparent wings occur in many groups of Lepidoptera, but particularly impressive and difficult to follow in flight are butterflies of the tropical American satyrine genus *Cithaerias* and species of many genera of Ithomiinae.

Counter-shading has the effect of flattening a rounded caterpillar, which consequently becomes less conspicuous. For example, the shadow produced by sunlight striking the caterpillar of the Eyed Hawk-moth, *Smerinthus ocellata* (Sphingidae), is cancelled out by the paler coloration of the downward facing surface, in this instance the dorsal surface as the caterpillar feeds upside down.

Flash-coloration is a device which reinforces crypsis by alternately startling and then deceiving

melpomene has been well documented, and in the moth families Arctiidae and Ctenuchidae there are several Müllerian complexes of chiefly yellow and black species in South America, for example between *Ormetica* and *Euplesia* species. In Italy the Burnet moth, *Zygaena ephialtes* (Zygaenidae), and the ctenuchid moth, *Syntomis phegea*, are apparently Müllerian partners or co-mimics.

Müllerian sound mimics have been found so far only in the Tiger-moths (Arctiidae), and Ctenuchidae, most of which can produce similarly pitched ultrasonic pulses from tymbal organs on each side of the thorax. Moths which have evolved an active night life have reduced their chances of being spotted by insectivorous birds, but are highly susceptible to bat predation. Colours cannot be used at night as warning signals of unpalatability and have been replaced by ultrasonic pulses which are audible to bats.

The large eye-like markings on the wings of many moths and butterflies frighten small birds by their resemblance to the eyes of large predators like owls and cats. This phenomenon is called eye-mimicry. Shock tactics are often used by moths having eye-spot markings: Emperor moths (Saturniidae), of the genus *Automeris*, for example the Io moth of North America and Mexico, keep their hind wing eye-spots hidden beneath the leaf-like forewings when at rest, but can suddenly reveal the eye-spot thus confusing small birds.

Cryptic adaptation or cryptic mimicry is one of the commonest protective strategies used by resting insectivorous birds. The partly day-flying Large Yellow Underwing, *Noctua pronuba* (Noctuidae), of temperate Europe and Asia has cryptically coloured forewings which completely cover the brightly coloured hind wings when the moth is at rest. When this species is disturbed and takes flight, the conspicuous colours of the hind wing will be revealed and a pursuing bird will be temporarily disconcerted. Sudden disappearance of the hind wing colours when the moth comes to rest again will further confuse the bird.

Feigning death, a characteristic of many Tiger-moths, (Arctiidae), is disputably either Müllerian or cryptic in nature, depending on whether the 'dead' moth is seen by the bird and identified as a dead unpalatable species (many Tiger- moths are highly poisonous) or whether, as often happens, the moth drops into the undergrowth and is hidden from view.

The small eye-spots at the other edge of the hind wings in some satyrines and Hairstreak butterflies, (Lycaenidae), are probably deflectors, unlike the large eye-spots mentioned earlier. Birds and other

The small eye-spots at the outer edge of the hind mistaking them for the butterfly's head; having survived a peck or bite at its wings, the butterfly is able to fly off. The effect of the false eyes of Hairstreaks is enhanced by the slender hind wing tails which resemble butterfly antennae and by the habit of some Hairstreaks of reversing their position immediately after landing so that the false head suddenly takes the place of the real head. A species of Cornelian butterfly, *Deudorix* (Lycaenidae), is said to run backwards, further deceiving an enemy as to which the real head end is.

A combination of deceptive measures is used by numerous species. Many tropical American Tiger-moths, for example, *Himerarctia,* have cryptically coloured forewings which when the moth is at rest cover the hind wings and abdomen, but when provoked the red hind wings and abdomen are revealed to warn an enemy that the moth is unpalatable. It is seemingly advantageous for these species to use crypsis as the first line of defence, unlike most warningly-coloured butterflies which openly advertise their colours throughout the day.

Toughness of the body is a characteristic of unpalatable Lepidoptera, but is just as important to the tropical Old World Moth-butterfly, *Liphyra brassolis* (Lycaenidae), whose caterpillars live and feed inside ants' nests and have a cuticle which is impregnable to the ants. Even tougher is the Puss-moth's cocoon, constructed of small chips of wood, which has the strength of a walnut shell. So tough is this cocoon that an emerging adult moth has to dissolve part of it away with caustic potash and then use part of the pupal case as a cutting tool before it can emerge.

Weaponry in the form of sharp spurs on the legs has been acquired by many large Hawk-moths (Sphingidae). The caterpillar and adults of many Tussock-moths, (Lymantriidae), some Emperor moths, (Saturniidae), for example, the South American genus *Hylesia*, some lasiocampid moths, Australian anthelid moths and the caterpillars of many families have developed hairs which are irritant to vertebrates, either chemically or physically. Irritant barbed hairs from the anal tuft of *Hylesia* moths cause the rash Carapito Itch in South America, and the dangerous caterpillars of the South American

Left : An example of shape and colour mimicry in the caterpillar of a European Plume moth Agdistis tamaricis *(Pterophoridae).*

Left : Colour mimicry in the Buff-tip moth, Phalera bucephala *(Notodontidae), which achieves a perfect match with the colour of the twig on which it is resting.*

saturniid *Lonomia achelous* produce an anti-coagulant which induces bleeding in man. The Puss-moth, *Cerura vinula* (Notodontidae), is a formidable adversary in its caterpillar stage which can eject a jet of formic acid from its thoracic glands and has long twirling red eversible tail filaments which may act as a deterrent to parasitic ichneumons. The osmeterium, a horned eversible mid-dorsal gland on the front of the thorax of Swallowtail caterpillars, (Papilionidae), secretes an acid in at least one species, the Old World Swallowtail, *Papilio machaon,* and may deter ants and parasitic wasps and flies.

One of the most effective ways of avoiding being eaten by a particular enemy is to ensure that encounters with it never occur, either by adopting a change of habitat or a change in the time of day or year when feeding or other activities take place. Escape into an aquatic environment by caterpillars of the South American Tiger-moth genus *Palustra* (Arctiidae), or the minute China-Mark moths,

Nymphula (Pyralidae), of northern temperate regions, has the effect of removing non-aquatic species from their list of potential enemies. Stem-boring, root-boring and leaf-mining caterpillars are similarly protected from most foliage-gleaning predators. Cossid moth caterpillars bore into living wood and are probably safe from most vertebrates other than woodpeckers and man. The day-flying life-style of agaristid, callidulid and castniid moths is almost certainly a secondary characteristic in evolutionary terms and may have resulted from predation pressures produced by insectivorous bats.

Flight itself releases insects from the attentions of terrestrial enemies, and from aerial predators too if the flight is fast enough—Morphos, for example, can out-manoeuvre most birds. Escape from one set of enemies, however, places an insect at risk with a completely new set; day-flying moths are prey for birds and aquatic caterpillars may be eaten by fish and carnivorous aquatic insects.

Above : The cryptic coloration of the common European Green Hairstreak butterfly, Callophrys rubi *(Lycaenidae), makes it merge with its surroundings when at rest on leaves.*

Friends and Enemies of Man

People who live in temperate or tropical climates generally take butterflies and moths for granted. They are familiar with them from childhood, and at some time or other may have kept caterpillars and watched their transformation into winged adults of an entirely different appearance. Lepidoptera are an integral part of the environment, and their aesthetic value is often underrated.

Many people take a lot of trouble to grow flowers and shrubs which are attractive to butterflies. In the tropics there is a wealth of such flowering shrubs, including many species of hibiscus, which attract some of the swallowtail and birdwing butterflies. Some of these plants will grow in sheltered situations in less warm climates. Among the most attractive are the buddleias, in particular *Buddleia davidii* which is sometimes known as the 'butterfly bush' and originates from China. Its blue, sweet-smelling flowers are visited in July and August by many of the common butterflies in the northern hemisphere, especially the vanessids, including the Red Admiral, *Vanessa atalanta*, Painted Lady, *Cynthia cardui*, Peacock, *Inachis io* and Small Tortoiseshell, *Aglais urticae*. What is often overlooked is that the caterpillars of these vanessid butterflies live on a certain stinging nettle (*Urtica dioica*). Thus a nettlebed in a sunny out-of-the-way spot in the garden can do a lot to help these butterflies to flourish. It does not matter if some of the nettles are cut down in the summer, as the new tender growth will be available for any caterpillars hatching later in the summer. Other garden plants which are especially attractive to butterflies include Michaelmas daisies (*Aster*), ice plant (*Sedum spectabile*), sweet rocket (*Hesperis*), sweet alyssum (*Lobularia*) and red valerian (*Centranthus*).

Many night-flying moths are attracted to flowers and may be searched out with a torch however, a red light should be used as this will not disturb them so much and consequently they are less likely to fly off.

Left, An adult Large White butterfly, Pieris brassicae *(Pieridae). Caterpillars of this butterfly often attack cabbages, sometimes causing extensive damage.*

It is possible to make home-made nectar for moths by brewing up equal amounts of brown sugar and molasses, or black treacle, mixed with some beer (stout) in an old saucepan. When the mixture has cooled it can be daubed on to tree trunks or posts with a paint brush. This should be done before dusk, preferably on a warm evening, and as darkness descends the moths will begin to arrive. This is known to collectors as sugaring, and the best results are usually obtained in late summer and early autumn.

The Silkworm or Cultivated Silkmoth, *Bombyx mori* (Bombycidae), must surely be the most useful moth in the world. It was the ingenious Chinese of ancient times who first reared silkworms and practised the art of spinning and weaving the threads from the silken cocoons. They guarded the process for more than 3000 years; they threatened death to anyone attempting to smuggle either the silkworm eggs or the seeds of the mulberry tree on which the caterpillars fed out of the country. Many centuries passed before sericulture spread to Japan and then to India, and many more centuries before it reached Europe. Not until about 950 AD did it spread to Sicily and other Mediterranean countries, after eggs and mulberry seeds had been smuggled out of China by two monks. It is believed that James I was the first to introduce sericulture in Britain, soon after he came to the throne, and he had mulberry plantations laid out near St. James's, in London. The silkworm venture was not a success and the plantations have long since disappeared, though a wizened mulberry tree remains in what is now the garden of Buckingham Palace. The tree is a black mulberry (*Morus niger*), and is labelled 'Planted 1609 when the Mulberry Garden was formed by James I'.

There have been other attempts at silkworm farming in Britain, and one which began as a modest venture has become famous. This is the Lullingstone Silk Farm which has supplied the silk for robes and dresses used on Royal occasions. It was started by Lady Hart Dyke in 1932, and until recently was at Lullingstone Castle in Kent, but has now been moved to Ayot House in Hertfordshire. Like many others Lady Hart Dyke kept silkworms as a hobby in her schooldays, and this inspired her to start her own silk farm. Its development and growth is a story of patience and determination, for silkworm farming has many hazards, not least of which is the difficulty of ensuring a constant supply of fresh mulberry leaves at the critical period when the larvae are feeding.

The caterpillars of many other moths also spin silken cocoons, but none surpasses the Silkworm which can spin more than a kilometre of unbroken thread of supreme quality. Species of saturniid moths belonging to the genera *Antheraea* and *Samia* are cultivated in India and the Far East and produce valuable silks, such as eri, muga, tasar or tussore and shantung.

Much less creditable to both eastern and western civilizations is the trade in exotic butterflies for purely decorative purposes. In South America the wings of countless thousands of morpho butterflies from the forests of the Amazon are used annually for making brooches and various other items for the tourist trade. Adults, caterpillars, chrysalids and even eggs are sold in large numbers to dealers and collectors. All over the world butterflies are used to adorn panels, pictures, table mats, ornamental caskets, ash trays and other trinkets, and rare and common species are collected indiscriminately for this purpose. Lepidopterists who collect specimens for study are sometimes blamed for the scarcity of some species but in fact are rarely responsible. Usually they take only a few selected specimens and they learn not only about them but also about the types of habitat inhabited by differe и species. This knowledge can be very useful when setting up nature reserves and similar protected areas.

A number of Lepidoptera are useful agents in the control of various weeds. The most outstanding example is the phycitine moth, *Cactoblastis cactorum*, which keeps in check the prickly-pear cactus (*Opuntia*) in Australia. Prickly-pear was introduced into Australia just before 1800, and in the following 100 years spread as an impenetrable gr и th over vast areas, including some of the best grazing land. It was

Below : Silkworm cocoons. European varieties of the Cultivated Silk moth, Bombyx mori (Bombycidae) have a single brood each year.

only brought under control in 1925 when the *Cactoblastis* from Argentina was reared in the laboratory and released in large numbers. The moth larvae tunnelled into the pulpy stems of the cactus causing them to rot, and within a few years the previously unassailable thickets of cacti were decimated and the land reclaimed for farming. Other countries have experimented with lepidopterous control of weeds with varying degrees of success. The *Cactoblastis* moth has been used against cacti in India, Sri Lanka and East Africa. The introduction of the Cinnabar moth, *Tyria jacobaeae* (Arctiidae), into New Zealand, Australia, Canada and the United States to control the alien European ragwort (*Senecio jacobaea*), has met with substantial success. Under favourable conditions the distinctive yellow-and-black ringed Cinnabar caterpillars strip the ragwort plants of leaves and flowers and can rid pasture land of this noxious weed. Since the European gorse (*Ulex europeus*) has become established in New Zealand and parts of the United States, including the Hawaiian Islands, several European

moths have been tested in attempts to control it. These include *Anarsia spartiella* (Gelechiidae), which feeds on the shoots, and *Cydia succedana* (Tortricidae), which feeds in the seed-pods.

Unfortunately the usefulness of some Lepidoptera is outweighed by the enormous amount of damage of one kind or another caused by Lepidoptera as a whole. In the adult stage they are rarely harmful, since those with tongues feed mainly on nectar. A few tropical noctuid moths, in particular *Othreis fullonia*, however, feed on the juice of fruits by piercing the skin with their proboscis, thereby opening the way to infection by fungus. In Thailand and Malaya some noctuids are attracted to the lachrymal secretions from the eyes of cattle and possibly carry infectious eye diseases from one animal to another. One of the noctuids, *Calpe eustrigata*, has recently been discovered to be a virtual vampire moth. It has a specially barbed proboscis which can be vibrated to drill and puncture the skin of mammals, especially cattle and deer, and then suck up the blood.

It is as caterpillars however, that Lepidoptera do such tremendous damage. The overwhelming majority feed off plants and few crops and forest trees are free from attack. Some species have adapted to a more sophisticated life-style and live on dry animal or vegetable matter, including substances which may have been preserved or perhaps processed by man.

The damage caused to crops and stored foods by these lepidopterous pests is immense, sometimes upsetting the economy of whole territories. Man has always tried to control them, and the methods used and the success rate have varied at different periods. During the last century, the battle against the most serious pests was largely a matter of gathering and destroying infested material. With the introduction of intensive methods of cultivation and the use of new chemical insecticides, the fight against these pests has taken on dramatic and hitherto unexpected features. Intensive cultivation itself favours the development of pests; for obviously an insect that lives on a particular foodplant will be able to multiply to a very great extent when its food supply is abundant. When an agricultural crop is being intensively cultivated, the conditions are ripe for a population explosion among the pest species.

With modern technology, the control of crop pests has been undertaken in a radical manner, initially almost entirely with the use of chemicals. Ever more lethal substances have been produced to poison and destroy insect and related pests, fungal diseases and weeds. The intention may have been to produce selective agents that acted against only the injurious species without harming the innocuous ones, but in practice this is rarely achieved and the long-term results have frequently been disastrous. Some chemicals used not only destroy every form of insect life with which they come into contact, but are dangerous to man himself. Furthermore, they have brought about changes in the ecological balance that were perhaps impossible to foresee, but which may now never be remedied.

An example which does not concern the Lepidoptera, but is of universal relevance is found in the rice growing areas of Italy which until relatively recently were infested with mosquitoes. To overcome the

Below : A silkworm, or larva, of the Cultivated Silk moth Bombyx mori (Bombycidae) feeds on mulberry leaves. The eggs of this commercially valuable moth and the seeds of its foodplant were originally brought to Europe from China.

inconvenience this caused, insecticides were used on a vast scale. The results seemed miraculous: it appeared that the insect pest had been eliminated. But it was no miracle—it was a tragedy. Together with the mosquitoes, many other water invertebrates also vanished; and their disappearance was followed by the decimation and near-disappearance of animals such as frogs, and certain species of fish that live mainly on water insects. The progressive disappearance of these creatures was due both to lack of food, and to the increasingly severe pollution of the environment by insecticides—which had been regarded as harmless for higher forms of life. As a direct result of the disappearance of the frogs and the fish, the herons and other waterbirds went too. Today, the commonest bird to be seen in the Italian rice-fields is the rook—a hardy and omnivorous bird which, now that it has no competitors, can invade the rice-fields the moment they have been sown, and make a clean sweep of the seed. As a crowning piece of irony, new forms of mosquitoes are appearing, which are ever more resistant to even huge dosages of insecticides. The original problem remains unsolved. As with mosquitoes, so too with Lepidoptera, the insecticides used against some lepidopterous pests have led to resistant strains being developed.

A sober and responsible approach to these problems has recently led scientists to turn to other methods of control, which limit, as far as possible, the expansion of harmful species while avoiding any further upset of the ecological balance. More and more interest is being shown in biological control. This term is applied to methods that make use of living creatures, of all kinds, to limit the damage caused to man's crops by insect pests. The methods are based on the realization that there are few animals that are not attacked by predators, parasites or pathogens at one phase or another of their lifecycle. Once the pest's natural enemies have been identified, the campaign passes to the laboratory, where the predator, parasite or pathogen is selected, bred and eventually dispersed in the area where the pest needs to be controlled. Biological control methods, therefore, depend on a careful study of the development of the insect to be controlled and a search for its natural enemies. The aim is not the extermination of a species, however harmful, but its reduction; not the total elimination of any damage, but the limitation of such damage to economically tolerable levels. Extermination of a species is very likely to lead to a state of severe imbalance, with consequences that cannot be foreseen but which are likely to be disastrous.

Among the most harmful moths in the Mediterranean region are the Pine Processionary, *Thaumetopoea pityocampa* (Thaumetopoeidae), and the Oak

Processionary, *T. processionea*. The caterpillars of the first construct pear-shaped or conical nests from which they emerge to feed on pine needles, while the caterpillars of the second live in nests on Turkey Oak (*Quercus cerris*) and various other oaks, also feeding on the foliage. Thaumetopoeid larvae generally live gregariously in large numbers within big nests, from which they emerge at night and set off in procession in search of food. They are known as processionary caterpillars because of their curious habit of following one another head to tail, forming a moving chain which may be as much as several metres long.

Processionary caterpillars are troublesome not only because of the damage they cause to trees, but more particularly because of their highly irritant hairs which, when they come in contact with human skin, can cause severe inflammation and itching. Both the living larvae and their shed skins or exuviae are harmful in this way; the hairs keep their irritant properties for a long time.

In the past various methods have been tried in attempts to eradicate processionary moths from some of the worst affected areas, but with little success. For a long time the only practical way of even partially limiting the numbers was to collect and destroy the larval nests. Fortunately a variety of insects and pathogens live on the eggs of these moths, their larvae or the adults, and some of them have already been used in attempts at biological control. Two of the insects tried are the ant *Formica rufa* and the carabid beetle *Calosoma sycophanta*.

Formica rufa is characterized by its light or dark red coloration (it is commonly called the red ant), with brown patches on the head and thorax. In southern Europe it is found mainly in coniferous woods, where it builds conspicuous hammock-shaped nests about a metre high and covered with pine and fir needles. The worker ants are avid predators of processionary caterpillars, which they will seek out on the highest branches of the pines. Nests of red ants were therefore taken from areas where they were plentiful and transported to forests where outbreaks of processionary moths occurred.

Calosoma sycophanta is a splendid beetle about 3 cm (1 in) long, with a blue head and thorax and green elytra or wing cases with a golden metallic sheen. It is one of the few flying species of the group, and is such a voracious eater that it might be very useful in the campaign against processionary moths which form its normal diet. Unfortunately it is not a very numerous species, and its use in the biological control of this species has not been successful.

Certain viruses, on the other hand, have been used against processionary moths with outstanding success. Viruses are the simplest form of micro-organism, and they live by penetrating into the interior of living cells in an active form; here they find all the chemical energy they need for their own metabolism and reproduction. They can also exist in an inert state outside living cells. In this inactive form, they can be used in the control of insects: once they have been introduced into the insect's body they multiply and spread within the body cells, eventually causing the death of their host.

One particular virus has proved of immense value in the control of processionary moths. It gives rise, in the caterpillars, to a disease in which polyhedral

Top left: Gregarious caterpillars of the Oak Processionary moth, Thaumetopoea processionea *(Thaumetopoeidae). Left: Caterpillars of the Pine Processionary moth,* T. pityocampa. *Both these species can cause severe damage to oaks and pines in southern Europe and the Mediterranean region.*

Right : The typical appearance of a lymantriid larva belonging to the genus Orgyia *(Lymantriidae). These caterpillars do not usually do great damage to the fruit trees on which they feed.*
Below : Caterpillars of the Peacock butterfly, Inachis io *(Nymphalidae), prefer the leaves of stinging nettles* (Urtica dioica) *but occasionally feed on other garden plants.*

crystalline inclusions found within the body cells causes a breakdown in the tissues and subsequent death. The most noted success attending the use of this organism was in a vast pine forest, over 500 hectares in extent, in the South of France. The trees in this forest were being devastated by hordes of processionary caterpillars and the whole forest appeared to be in jeopardy. Processionary caterpillars were bred in the laboratory and infected with the virus. Then, when dead, they were dried and ground to powder, and a suspension of this powder in water was sprayed over the forest from specially equipped helicopters. In no time the disease spread among the caterpillar pests and reduced the population to a more tolerable level.

Viruses have also been used against other lepidopterous plant pests. A classic example is that of the Small White and Large White butterflies, *Pieris rapae* and *P. brassicae* (Pieridae), whose caterpillars are extremely destructive to cabbages. In certain regions of the world it was observed that many *Pieris* caterpillars were dying of a highly contagious virus infection which left them flaccid. Infection occurs by contact with other infected insects, or by ingestion of virus capsules spread over the foliage by rain. Viruses are carried by winds and by insects themselves, which explains their diffusion throughout the world. After a great deal of laboratory and field research on the reproduction, pathogenicity and culture of the virus, it was found that a water suspension of a few, ground-up dead caterpillars produced an effective spray for disseminating the disease. This granulosis virus can only be cultured on

tissues of Lepidoptera and is only lethal to this group and is harmless to all others.

Bacteria, as well as viruses, can be put to effective use in biological control campaigns; and their action is no less specific than that of viruses. *Bacillus thuringensis*, for example, produces a toxin which is highly effective against lepidopterous caterpillars specifically, while being harmless not only to vertebrates but to other insects as well. This is a vital consideration, since otherwise it would be highly risky to use the bacterium because of the danger of upsetting the ecosystem. Not even the larvae of parasitic Hymenoptera that may be inhabiting the bodies of infected caterpillars suffer any ill effect from their host's disease. Cultures of *B. thuringensis* have been prepared and marketed commercially as dusts or sprays, and have proved effective in the disinfestation of a wide variety of agricultural crops. Recently, Soviet scientists have shown that the effectiveness of this bacterium against yponomeutid caterpillars that attack fruit trees can be greatly increased by combining it with minute doses of DDT.

A critical time when an effective attack may be made on an insect pest is the reproduction or mating stage. If this process can be upset, the pest may be automatically controlled by the reduction of its progeny. Essentially, this involves sterilization of the males of the species to be controlled. These measures have so far mainly been used successfully against Diptera. For example, male pupae of the dipterous species *Callitroga hominivorax* (a parasite that inhabits the muscle tissue of domestic cattle) can be sterilized by irradiation; these infertile individuals then compete in the normal manner with fertile flies in the processes of courtship and mating, with the result that a percentage of the eggs that are laid are infertile; the number of adults alive during the next reproductive season is thus automatically reduced. If the process is repeated for a few seasons in succession, the results are far better than any that could be obtained with insecticides. The same has been done with other species of Diptera, such as the Mediterranean Fruit-fly, *Ceratitis capitata,* a pest of citrus fruits, peaches and apricots; and with certain species

Above: A male Brimstone butterfly, Gonepteryx rhamni *(Pieridae). Caterpillars of this common European species live on buckthorn* (Rhamnus) *and do not damage cultivated plants.*

of culicine mosquitoes. It is possible that identical techniques may eventually be applied against the multitude of lepidopterous pests which damage fruit trees and crops.

A similar avenue of control which is being investigated is the manufacture of synthetic sex attractants or pheromones. These have already shown promising results with some species, such as the Gypsy Moth, *Lymantria dispar* (Lymantriidae), in North America. They are based on the presence of scent organs in the female for attracting the male. Such organs are present in many Lepidoptera, and are usually on the last segments of the abdomen, either as gland cells bearing hairs or as tufts of modified scales, and function when the female has emerged from the pupa and is awaiting fertilization. So powerful are these scents in some species of eggar moths that they can be detected over a distance of ten kilometres or more. The pheromones produced by these and many other Lepidoptera are usually characteristic of the species. By liberal use of the scent the males can either be confused so that they cannot find the females, or can

be caught in traps which have been baited with the pheromone.

Although they have not yet been employed for biological control of Lepidoptera, many kinds of fungi are parasitic on insects. The muscardine diseases are fungal growths which envelope the insect in a white or green fungal shroud, and two Lepidoptera which are commonly attacked are the Silkworm, *Bombyx mori*, and the Codling Moth, *Cydia pomonella* (Tortricidae). The fungus *Cordyceps* usually attacks subterranean larvae, and produces bizarre effects in some of the large underground caterpillars of the Swift moths (Hepialidae), in Australia and New Zealand. The fungus fills the whole body of the caterpillar and then produces a stem-like stroma which pushes upwards to the surface of the soil and may be 20 cm (8 ins) long.

One of the most promising approaches to pest control is the combination of chemical and biological methods. This has become known as integrated control and when it can be applied it is generally found to be much less harmful to the environment.

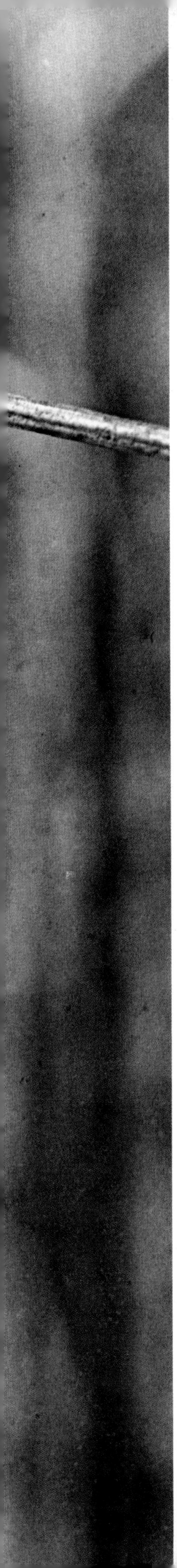

The Classification of Lepidoptera

In the twentieth century it has become traditional among collectors of Lepidoptera to divide the order into two major groups, the Macrolepidoptera (butterflies and larger moths) and the Microlepidoptera (smaller moths). The Macrolepidoptera are then further divided into Rhopalocera (butterflies) and Heterocera (moths); the former usually being distinguished by the presence of clubbed antennae, while the latter have antennae of various forms.

A more precise classification favoured by some lepidopterists is the simple division of the Lepidoptera into two suborders, namely Homoneura and Heteroneura. Homoneura comprises only a few primitive families in which the wing venation is similar in both the forewing and hindwing; the wing-coupling apparatus consists of a jugum and frenulum or a jugum alone, the segmentation of the abdomen is primitive and the female usually has only one genital aperture. Heteroneura comprises all the butterflies and the majority of the moths in which the forewing venation differs from that of the hindwing, the wing-coupling apparatus normally includes a frenulum and the female usually has two genital apertures.

However, while one purpose of classification is convenience, a more important one is to show relationships. With advancing knowledge the Homoneura-Heteroneura scheme has now been superseded by a more realistic division of the Lepidoptera into four suborders: Zeugloptera, Dacnonypha, Monotrysia and Ditrysia, whose characteristics and phylogenetic relationships are shown in the diagram.

In spite of the mass of work that has been carried out worldwide on the classification of the Lepidoptera, the subject is still a long way from any definitive order. In the case of some families, the prevailing taxonomic chaos makes it impossible, at least for the time being, to attempt any reliable distinction between some of the component subfamilies, genera and species.

Left: A freshly emerged Swallowtail butterfly, Papilio machaon *(Papilionidae). The wing pattern is very pronounced, and the hindwings are extended in a pair of tails which are characteristic of the papilionids.*

LEPIDOPTERA (Scale-wing insects)

A phylogenetic diagram outlining some characteristics of the subordinal groups of the Lepidoptera

archaic moths
Suborder: ZEUGLOPTERA
Adults with mandibles for chewing; fore- and hindwing with similar venation, with fibulate coupling; female with single sex opening (cloaca)
Micropterigoidea

primitive moths
Suborder: DACNONYPHA
Adults with vestigial mandibles or rudimentary tongue (proboscis); fore- and hindwing with similar venation, with fibulate coupling; female with single sex opening (cloaca)
Eriocranioidea

primitive moths
Suborder: MONOTRYSIA
Adults without mandibles, usually with short proboscis; fore- and hindwing venation similar, or reduced in hindwing; female with one or two genital openings on segments 9–10.
Nepticuloidea
Incurvarioidea
Hepialoidea (female genitalia exoporian, intermediate between this suborder and Ditrysia)

advanced moths and butterflies
Suborder: DITRYSIA
Adults with specialized sucking mouth-parts, i.e. proboscis, but this sometimes degenerates; venation of fore- and hindwing dissimilar; wing coupling frenulate or amplexiform

Butterflies (Rhopalocera)
Antennae clubbed; hindwings without frenulum; wing coupling amplexiform

Moths (Heterocera)
Antennae of various forms but rarely clubbed; hindwing usually with frenulum; wing coupling usually frenulate but sometimes amplexiform, e.g. Bombycoidea.

Suborder ZEUGLOPTERA
Superfamily Micropterigoidea
Family Micropterigidae (Micropterigids)

These are generally regarded as archaic Lepidoptera and the most primitive of the moths. There are about 80 known species of this family which are dispersed in the temperate regions of the world. In Europe and North America the family is represented by the genus *Micropterix*, containing about 50 species, and in Australia and New Zealand by the smaller genus *Sabatinca*. The adults are diurnal and mostly have a bronzy metallic coloration and a wingspan of 7 to 15 mm (0.3–0.6 ins). They are unusual among the Lepidoptera in having mandibles which they use for chewing; they feed on the pollen of herbaceous plants, trees and shrubs.

Suborder DACNONYPHA
Superfamily Eriocranioidea
Family Eriocraniidae (Eriocraniids)

This is another small family of primitive moths, but is apparently restricted to the Holarctic region. The adults have mouth-parts intermediate between those of the Micropterigids and the more usual haustellate, or sucking type seen in most Lepidoptera. They are diurnal and fly in sunshine, sometimes in swarms and are attracted to birch trees.

Eriocraniid larvae are leaf miners, and their mines are easily recognized by the long, intertwining threads of frass deposited inside. Until recently Eriocraniidae were believed to occur in Australia, but careful research has shown that the Australian species represent two separate families. These have been named Agathiphagidae and Lophocoronidae.

Suborder MONOTRYSIA
Superfamily Hepialoidea
Family Hepialidae (Swift Moths, Ghost Moths, Porina Moths)

This is a world-wide family of comparatively large, robust moths, common in grassy, temperate regions and particularly numerous in Australia and New Zealand. Some of the larger species can have a wingspan of 15 to 20 cm (6–8 ins) and are powerful fliers, hence the name Swift moths. Their livery varies from sombre browns and creams to brilliant greens, sometimes embellished with silvery or mother-of-pearl markings. The largest European hepialid is the Ghost moth, *Hepialus humuli*, with a wingspan of 7 cm (3 ins); its larva is subterranean and feeds on the roots of grasses, hops and nettles. The females of some hepialids scatter the eggs as

Left: A micro-moth, belonging to the genus Simaethis, (Glyphipterigidae), whose shape and colouring are very similar to those of the Tortricidae.

they fly, dropping them at random on the chance that they come to rest near a suitable food plant for the larva. In Australia and New Zealand, besides the many subterranean larvae which damage pasture, particularly those of the Porina moths of the genera *Wiseana* and *Oxycanus*, larvae of some of the larger species tunnel in the living wood of the eucalyptus tree and also in other trees and vines, often weakening and harming them.

Superfamily Nepticuloidea
Family Nepticulidae (Nepticulids, Pygmy Moths)

These minute micro-moths include the smallest living lepidopteran known, which is found in Britain and Europe. It is *Stigmella acetosae*; its larva mines leaves of the sorrel (*Rumex acetosa*). Despite their small size, nepticulids are found in all parts of the world from sea-level up to 3500 m (11,460 ft). The adults sometimes have drab-coloured wings, but more often their wings are banded in glittering metallic gold or silver.

Nepticulid larvae are mostly leaf miners, tunnelling in the leaf parenchyma, but some mine in the epidermis of stems or in buds. When the mining larva reaches maturity it usually abandons its leaf and descends to the ground, where it pupates within a minute silken cocoon.

Family Tischeriidae (Tischeriids)

These are very small and inconspicuous Microlepidoptera; the majority of species occur in America, only about eleven being known from Europe, five of them in Britain. No species of this family has so far been found in Australia or New Zealand. The larvae are leaf miners, usually making characteristic blotches in the leaves and ejecting all their frass (excrement) through a hole in the leaf epidermis. Most of the species are seldom very common, but occasionally the European *Tischeria gaunacella* is sufficiently plentiful to cause marked disfiguration to ornamental cherries (*Prunus*), having two or three successive generations in the course of a year when conditions are favourable.

Superfamily Incurvarioidea
Family Incurvariidae (Incurvariids, Adelids, Yucca Moths)

Three subfamilies are recognized: Incurvariinae and Adelinae, both of which have a worldwide distribution, and Prodoxinae which is American. In the larval state, the

Incurvariidae are leaf miners sometimes they live on the seeds inside fruit or tunnel in young shoots; some later cut out flattened cases which they use as portable shelters and also for pupation. The species *Incurvaria koermeriella* is quite widespread in Europe, it mines in the leaves of limes and beeches until fully grown, then descends to the ground and constructs a case from fragments of plant materials, within which it overwinters before pupating.

The Adelinae are distinctive by their remarkably long, filiform antennae, those of the male being much longer than of the female and sometimes four or more times the length of the forewing. One of the commonest European species is *Adela viridella* whose forewings are a bright brassy-green.

These elegant insects like to fly in sunshine and often assemble in dancing swarms around oaks, when they appear as scintillating flashes of light against the dark background of the vegetation.

The subfamily Prodoxinae is a New World specialization, and have evolved as borers in agavoid plants of the genus *Yucca* in particular. The plants are dependent on the moths for pollination. The female Yucca moth appears on the scene when the plants begin to flower and gathers the pollen from the anthers of different flowers with the aid of her modified mouth-parts. The pollen is compressed into a ball between the base of the forelegs and the neck and in this manner is transported and deposited on a stigma or forced down the stigmatic tube of the flower. The female lays her eggs in the pistil next to the ovules of the fertilized plant, so that the larvae can feed on the developing Yucca seeds.

A fascinating fact is that Yucca plants fertilized by moths of the genus *Tegeticula* are also the food of a closely related group of moths of the genus *Prodoxus*, whose larvae live on the stems and the flesh of the fruits of the plants. The survival of the latter group of moths is dependent on the continuing success of the *Tegeticula* moths in pollinating the Yuccas.

Family Heliozelidae

This is a small family of about 100 species, with representatives in most regions except New Zealand. The moths are very small, having a wingspan of only a few millimetres, and often have metallic scales; they fly in sunshine and bask on flowers. The larvae are leaf miners. Where vines are cultivated in Europe it is not difficult to find *Holocascista rivillei*, a micro-moth with a wingspan of about 4 mm (0.2 ins). Its larva bores a long twisting mine in the vine leaves, which finally terminates in a broad central patch. When fully grown the larva cuts out two symmetrical oval segments from the upper and lower membranes of the leaf mine and from these constructs a portable case in which it descends to the ground and pupates. When environmental conditions are favourable large numbers of *H. rivillei* larvae may develop and the resulting excessive damage to the leaves may debilitate the vines. In Australia a similar species, *Heliozela prodela*, mines in the young leaves of eucalyptus.

Suborder DITRYSIA
Superfamily Cossoidea
Family Cossidae (Goat Moths, Carpenter Moths, Wood Moths)

These are typically large, stout-bodied moths of rather sombre greyish appearance. *Xyleutes boisduvali*, an Australian species, may attain a wingspan of 24 cm (9 ins), with a body diameter of 2.5 cm (1 in). Its larva bores in the trunks of the eucalyptus and lives two or three years. The 'Witchety grub' of Aboriginal folklore is the larva of a cossid living in the roots of the acacia.

In Europe the Cossidae are represented by about 10 species, some of which are of considerable economic importance as agricultural pests. One of the commonest is the Goat or Carpenter moth, *Cossus cossus*, which has a wingspan of nearly 4 cm (1.5 ins). The females lay their eggs in the crevices of the bark of mature trees such as elm, ash, poplar and lime. When the larvae hatch out they start life under the bark and then burrow into the living wood, boring long tunnels in an upward direction. Mature larvae

may be 10 cm (4 ins) long and have a rather attractive wine-red and cream coloration, but they repel close examination by emitting a rank goat-like odour which is extremely unpleasant.

Another common European cossid is the Leopard moth, *Zeuzera pyrina*. This lays its eggs in the bark of woody plants, including pear, apple, birch, cherry, lilac, vines and laurel. It is very different from the Goat moth in appearance, being white with steel-blue spots on its body and wings. It is in fact generally placed in a separate subfamily, the Zeuzerinae, together with a number of other species of similar appearance. The female Leopard Moth lays her eggs at the base of the leaf stem, on the buds, or in the folds of crevices of the bark of the branches. Newly emerged larvae spend a short time under the bark and then burrow into the living wood, making ever more spacious galleries. They sometimes cause severe damage, interrupting sap flow with consequent failure of fruit to develop and deterioration of the condition of the plant as a whole.

Moths of the genus *Zygaena*, known as Burnets, are typically black with red or yellow patches on the wings; those of the genus *Procris*, known as Foresters, are typically a uniform iridescent green or blue. Zygaenid larvae are generally short and stubby, with numerous bristly tubercles. They usually pupate in a parchment cocoon attached to the food plant or to an object nearby.

The genus *Zygaena* is almost confined to the Palaearctic region and only a few species occur outside. Some of the species are very common in Britain and Europe and have a strong tendency to congregate and live in colonies. Perhaps because of intensive inbreeding, these colonies often show slight differences and experts have divided them into an almost infinite number of subspecies. Their conspicuous red and black coloration warns potential predators that they are distasteful.

The taxonomy of the genus *Procris* is also very complex, and confident identification of some species is only possible if the genitalic structures are dissected.

Above : The Goat or Carpenter moth, Cossus cossus *(Cossidae), a widely distributed species in Britain, continental Europe, North Africa, and found even in Siberia. The larva tunnels in the trunks of deciduous trees and produces a strong and unpleasant odour like that of a billy-goat.*

Superfamily Zygaenoidea
Family Zygaenidae (Burnets, Foresters)

A worldwide family consisting of mostly handsome, medium-sized, day-flying moths, with long narrow wings, often marked with bold patterns and metallic colours.

Family Limacodidae (Slug-caterpillar Moths, Cup Moths, Nettle-grub Moths)

This mainly tropical and subtropical family has two representatives in Britain and Europe. One is the Festoon moth, *Apoda avellana*, the larva of which feeds on oak and beech,

and the other is the Triangle moth, *Heterogenea asella*, a drably coloured moth with triangular-shaped forewings. Its larva feeds on beech as well as other trees, and it is reported to have been introduced to North America. Limacodid larvae are short and thick-bodied, and bear a vague resemblance to slugs, hence the family name Limacodidae.

Family Heterogynidae

A small family of rather primitive moths found in Europe and North Africa. *Heterogynis pennella*, which occurs in southern Europe, exhibits marked sexual dimorphism, the male being winged while the female is not only wingless but spends almost the whole of her life withdrawn inside the pupal case.

Superfamily Tineoidea
Family Tineidae (Clothes Moths, Skin Moths)

This vast family of generally small-sized moths includes some 2000 described species. It is divided into a number of not very dissimilar subfamilies. The adults are characterized by their rough-scaled heads, bristly labial palpi, short or rudimentary proboscis, and a habit of resting with their wings steeply folded roof-like over the abdomen. The

larvae have a wide variety of feeding habits: they may live on dead or decaying animal and vegetable matter, particularly fungi, or on living plants; some live in the nests of birds, feeding on feathers and debris, or in nests of ants and termites and the habitation of other animals.

Many tineids do a great deal of damage, attacking foodstuffs, clothing, household furnishings, and industrial products of various kinds. The internationally notorious Common Clothes moth, *Tineola bisselliella*, belongs to this family, but the term clothes moth is commonly applied to a number of tineid species, including the Case-bearing Clothes moth, *Tinea pellionella*, and the Tapestry moth, *Trichophaga tapetzella*; these all live on wool, hair, furs, feathers and spin silken tubes or little portable cases. One of the species which attacks stored food products, the Corn moth, *Nemapogon granella*, is found in all the temperate regions of the world and causes great damage in granaries and food warehouses.

Family Psychidae (Bag-worms, Psychids)

This is a worldwide family of small to medium-sized primitive but highly modified and distinctive moths. The adults have no functional tongue or mouth-parts and do not feed during their brief life; the females are usually wingless and never leave the case or bag in which they developed from

Above : The moth Euplocamus anthracinalis (Tineidae), one of the handsomest European species, whose caterpillars live on rotten wood of oak, beech and hawthorn. Tineid moths often live on wool, hair, fungus or decaying organic matter.

the larval stage. Indeed, the females may show varying extents of physical involution, including the loss of eyes, antennae and even legs. The larvae of Psychidae build themselves highly elaborate cases which are characteristic of their particular species. The cases are made from silk secreted by the labial glands, and may have an almost smooth surface or may be adorned with fragments of leaves, twigs, lichen, grains of sand or tiny shells. Some resemble little bundles of twigs and are reminiscent of the larval cases of aquatic caddis flies. The larva moves its entire case in search of food, never leaving the case but feeding through an opening at the front or mouth and ejecting its frass through an opening in the tail end of the case. As it grows it enlarges its case accordingly. Psychid larvae are essentially herbivores and do not have enormous appetites like the larvae of most other families; when their preferred food is scarce they generally adapt to whatever is available. A few are carnivores and that is to say that their survival depends on their eating other insects.

In Europe there are about 50 species of psychids, none of which is a true pest. *Pachythelia villosella* is a common inhabitant of meadows; it has quite a large case, covered with dry leaves and bristling with long backward-projecting stalks. Species belonging to the genus *Cochliotheca* have characteristic helical-shaped cases constructed of silk and sand.

The family Psychidae may be divided into several sub-families, including the Solenobiinae and Taleporiinae both of which are particularly interesting. The first is based on the genus *Solenobia* whose species are mostly European, with stragglers in the eastern Soviet Union and North America, and the second on the genus *Taleporia* which is represented in all regions. In both genera the females are apterous; the larvae feed on lichens, living in elongated portable cases which are often covered with bits of lichen or grains of sand. The genus *Solenobia* is important for the studies on parthenogenesis carried out on it: fertile eggs may be laid by unimpregnated females for several generations in succession before a male is required or appears. The genus *Taleporia* has been used in interesting experiments on the influence of environmental conditions on the sex ratio. For instance, when the species *T. tubulosa* is bred at temperatures below 18°C (64°F), mainly females are produced, while at temperatures between 30° and 37°C (86–99°F), the balance shifts in favour of males.

Family Lyonetiidae (Lyonetiids)

This family of micro-moths comprises 1000 species or more and is found worldwide; it is especially numerous in the Australasian region. The classification of this group is still poorly known. Adults of many of the species are characterized by the upturned or downturned tips of the forewings, while others, for example, *Lyonetia*, have normal flat wings which are characteristically shining white with delicate linear markings. The larvae are usually miners and make blotch or serpentine mines in the leaves, or mine in the bark, but the more primitive forms are refuse feeders.

The genus *Lyonetia* is represented in most temperate and tropical regions except New Zealand. It contains the common European Apple Leaf-miner, *L. clerkella*, whose larva makes long irregular galleries in the leaves of apple, hawthorn, plum and birch. A related species, *L. latistrigella*, occurs in North America and mines in the leaves of rhododendrons. The genus *Leucoptera*, which is likewise widely distributed but is not found in New Zealand, includes a number of economically important species.

Family Gracillariidae (Gracillariids, Leaf Blotch Miners)

A family of small elegant moths which is found worldwide. It generally has long, narrow forewings and very slender pointed hindwings, with antennae distinctly longer than the forewing and with long upcurved labial palpi. The moths have a characteristic resting posture, folding the wings tightly along the body and raising the head and forepart of the body steeply on long ornamented fore and middle legs.

Below: The Currant Clearwing moth, Synanthedon tipuliformis *(Sesiidae), may be easily mistaken for an ichneumon wasp. Originating from Europe, this moth has been introduced into North America, Australia and New Zealand.*

Superfamily Yponomeutoidea
Family Sesiidae (Clearwings, Squash Borers)

This remarkable family comprises small to medium-sized moths which are found throughout the world and are remarkable for their mimicry of different species of bees and wasps. Their wings are usually only partially scaled and the abdomen ringed with black and yellow bands—their resemblance to Hymenoptera is so good that they are often mistaken for wasps or hornets. Some carry the imitation even further and can make a buzzing noise.

Sesiid larvae live internally, tunnelling in the bark, trunks, branches and roots of a variety of wild and cultivated plants, and pupate within their burrows. The pupae have rows of backward-curved spines on their backs, and when the time comes for the adult to emerge the pupa uses these to work its way along the burrow to an exit hole earlier prepared by the larva. One of the best known species is the Currant Clearwing, *Synanthedon tipuliformis*, which is often a nuisance to growers, boring into the main stem and branches of gooseberry and currants (*Ribes*), feeding on the pith and working its way down the stem. Originating from Europe, this species has been introduced into North America, Australia and New Zealand. The adult has a wingspan of about 2 cm (0.7 ins); the male has four yellow belts on the body, the female three. One of the largest sesiids found in Europe and North America is the Hornet Clearwing, *Sesia apiformis*, which as its name implies mimics the Hornet Wasp, but lacks its sting. Sesiid moths are very fast fliers and extremely difficult to catch. Thus many species are rare in collections and very little at all is known about the life histories of those species living in the tropics.

Family Glyphipterigidae

A large family of micro-moths, comprising 1000 species or more and most numerous in the tropics. The adults are usually day-fliers and have bright metallic or orange wing markings. The genus *Glyphipterix* contains over 200 species distributed throughout the world. The larvae feed on seeds or bore into shoots or flower-stems.

The widely distributed genus *Anthophila*, includes a number of species found in the warmer parts of Europe and in the tropics whose larvae feed on the foliage of fig trees. The moths of this genus have comparatively broad wings and when at rest these are raised and curled. A common species in Europe is *A. fabriciana*, whose whitish larva feeds in a small web on nettle leaves (*Urtica*).

Family Douglasiidae

A small family of worldwide distribution comprising only about 20 species. In general appearance the adults might be mistaken for elachistids but usually have narrower and more pointed wings. The larvae mostly live in the stems and flowers of borage. Only two species are found in Britain, both on viper's bugloss (*Echium vulgare*).

Family Yponomeutidae (Ermine Moths)

This is a vast family of micro-moths and is usually divided into several subfamilies, including the Plutellinae, Acrolepiinae and Argyresthiinae. The species are most numerous in tropical regions, where the largest and most brightly coloured forms occur. The typical small Ermine moths include the Orchard Ermine, *Yponomeuta padella*, which is common throughout Europe, including Britain. It has a wingspan of about 2 cm (0.7 ins) and is easily recognizable by its white forewings sharply marked with numerous little black spots. The caterpillars feed on the foliage of apple, plum and hawthorn, living gregariously in webs. There is only a single generation each year, and the females lay small batches of eggs on the twigs in June and July. The larvae hatch about 10 days later and immediately prepare for hibernation, spending the autumn and winter sheltering in little dome-shaped refuges composed of their own eggshells. In spring they leave this refuge and set off in groups to mine the leaves of their hostplant. At a certain point in their development they cease mining and begin chewing the

Above : The Hornet moth, Sesia apiformis (Sesiidae), showing a strong but entirely superficial resemblance to a wasp or hornet. This species occurs in Europe and North America.

The larvae are mostly leaf miners, making inflated bubble-like mines, or they roll the leaves into conical chambers in which they live. One of the most attractive species is the Azalea moth, *Caloptilia azaleella*, which has purple-black forewings with a yellow patch along the front margin. This tiny moth, which has a wingspan of about 15 mm (0.5 ins) is a pest on azaleas. It is believed to originate from the Far East and has become established in Europe, including southern England, and North America. The larva mines the azalea leaves and when nearly fully grown emerges from the mine to curl the tips of the leaves back with silk. When conditions are favourable it has been known to defoliate whole bushes.

Family Phyllocnistidae

Minute moths which mostly have shining white forewings with delicate yellowish patterns. Over 50 species are known from temperate and tropical regions, but none from New Zealand. The larvae have no legs. In some species, such as the common European *Phyllocnistis saligna* that lives on the willow (*Salix*), the larva begins by mining a leaf but soon abandons it, tunnelling down the leaf stem and into the bark of the twig and through it until it reaches another leaf into which it passes and then completes its development.

leaves from outside, wrapping the leaves together in a web of silk.

Y. padella is a member of a species complex in which the adults are indistinguishable but whose larvae are generally specific to particular foodplants. An economically important yponomeutid in countries of the Mediterranean region is the moth, *Prays oleellus*, a little grey moth with small blackish patches on the forewings. It can be a serious pest, its larva feeding on the leaves, flowers and fruit of olive trees. Generally there are three generations of this species in a year, the first living on the leaves, the second on the flowers and the third on the fruit. The damage done to the leaves is insignificant, but that to the flowers can be serious and that to the fruit catastrophic which may prove costly in terms of loss to the grower.

The subfamily Plutellinae contains about 200 species, occurring in all regions. One of the best-known and most universally distributed species of the Lepidoptera despite its small size, is the Diamond-back moth, *Plutella xylostella*. It is a pest of cabbages, cauliflowers and other garden crucifers. The fully grown larva is about 8 mm (0.3 ins) long, pale green in colour, and skeletonizes the leaves of the plants; it pupates in a very characteristic flimsy open-mesh cocoon attached to the leaves. Another harmful species in Europe, usually placed in the separate subfamily Acrolepiinae, is the Leek moth, *Acrolepiopsis assectella*, whose larvae skeletonize the leaves of leek, garlic and onion (*Allium*).

Family Epermeniidae

These are small, narrow-winged micro-moths, numbering less than 100 species but found in most parts of the world except New Zealand. They are generally recognizable by the presence of scale-tufts on the hind margin of the forewing and stiff bristles on the tibia of the hind leg. The larvae mine in leaves in their early instars, especially of Umbelliferae, afterwards feeding externally in a slight web.

Superfamily Gelechioidea
Family Coleophoridae (Casebearers)

This is a worldwide family of delicate, little moths with long fringes to the wings, whose life-style is characterized by the larval habit of making highly intricate, portable cases in which they live and carry around. The caterpillars usually feed on seeds and leaves, and as they grow bigger and enlarge their cases the species can usually be identified by the style and construction of case. Several hundred species are known from Europe and North America, but very few from the tropics and the southern hemisphere. The species form numerous compact little species-groups or subgenera which are usually associated with a particular family of plants. These are generally considered to form the one large genus *Coleophora*, which is the most practical method of classifying

Below : A common European pyralid moth, Endotricha flammealis, *in its characteristic resting posture. The larvae live on decaying and fallen leaves.*

them. One species of this group is associated with clover and lucerne, and is of economic importance. It includes the Small Clover Case-bearer, *C. alcyonipennella*, which, like the other members of this particular group has metallic green-coloured forewings and, in the male, thickly scaled antennae. The larva hollows out the seeds, particularly of white clover, greatly reducing seed yield. It is a common species in Europe and has been introduced, probably with cattle fodder to Australia and New Zealand, and more recently has been reported from North America.

Family Elachistidae

This family comprises several hundred small and mostly obscure micro-moths distributed throughout the world and even reaching remote subantarctic islands. The larvae mostly mine the leaves of grasses and sedges. One of the commonest of the European species, *Elachista argentella*, is exceptional in having almost pure silvery white forewings without markings.

Family Oecophoridae

A worldwide family comprising several thousand species, a high proportion of them endemic to Australia, where many live in the eucalyptus forests and play an important role in the ecosystem, feeding on the very dry litter peculiar to these forests. Many of the moths are colourful and have yellow, orange or pink coloration; a distinctive feature is their very long, curved labial palpi. In the northern hemisphere one of the best known genera is *Depressaria*, which contains many species whose adults hibernate. The moths of this genus have a characteristic flattened or depressed form, an adaption to enable them to squeeze their way into thatch and crevices. Many species of this genus and of the closely related genus *Agonopterix* are attached to umbelliferous plants and Compositae, the larvae living in the flowers and seeds or boring into the stems. In southern Europe *D. erinaceella* lives on artichokes, the larva attacking first the leaves and then the principal veins and finally the inside of the head itself.

Several oecophorids are household pests, including the Brown House moth, *Hofmannophila pseudospretella*, which is a cosmopolitan species probably originating from Asia. Its larva is omnivorous and lives on almost anything chewable, from hard seeds to flour, cork, carpets and soft furnishings and even polythene and other plastics, though whether it derives any nourishment from the latter materials is doubtful. Another widespread domestic species is the White-shouldered House Moth, which has a conspicuous white head and 'shoulders' and grey mixed with black forewings. Although common in households and warehouses it is rarely a pest in the sense that it destroys food or clothing, but tends to live on accumulations of debris under skirtings and floorboards and in cupboards.

Family Ethmiidae

A small family of medium-sized micro-moths which is most numerous in the tropics. Some of the adults are superficially similar to the small Ermine moths, having white forewings with black spots, but many are more colourful and have yellow hindwings. The larvae generally feed externally in a slight web and are especially attached to Boraginaceae.

Family Gelechiidae

This family is of worldwide distribution and consists of several thousand species of micro-moths which may be grouped into a number of not always clearly definable subfamilies. In general, however, gelechiid moths may be recognized by their characteristically shaped hindwing, which is trapezoidal and has a produced or pointed tip and a concave outer margin. The larvae mostly feed on seeds or among shoots, but some are miners, boring tunnels in stems or tubers, and a number cause galls. Many species are important agricultural pests. One is the Peach Twig-borer, *Anarsia lineatella*, a Palaearctic species that has been introduced into North America and Australia. The immature larvae spend the winter in refuges excavated in the branches of the host tree; these shelters are lined with silk and connected with the outside by a silken tube incorporating the larva's own excrement. In spring the larva emerges from this retreat and burrows into the young shoots, causing them to wither.

A gelechiid that is found wherever cotton is cultivated is the Pink Boll-worm, *Pectinophora gossypiella*, so-called because of the pinkish colour of the larva. It is a serious pest in many cotton growing countries, and is rated among the top 10 most serious insect pests in the world. The larvae penetrate into the cotton bolls to feed on the seeds, and in the process damage the cotton fibres. In the United States and other cotton producing regions outbreaks of this moth have been known to ruin more than half the year's crop.

The Potato Tuber moth, *Phthorimaea operculella*, is another worldwide species of major economic importance. This species breeds continuously in warm climates, with several generations in a year, and besides attacking growing potato tubers or stored potato tubers it will feed on tomatoes and tobacco plants. A closely related species, the Beet moth, *Scrobipalpa ocellatella*, is a pest of sugar beet in central Europe and the Mediterranean countries. The females lay their eggs on the leaves, and on hatching the young larvae bore into them and then down the stalk and into the root.

Below : A Grass moth of the genus Crambus *(Pyralidae). These moths are plentiful in Europe and North America and the caterpillars feed among stems and roots of grasses or on moss.*

One of the most serious pests of stored grain of all kinds is the Angoumois Grain moth, *Sitotroga cerealella*, which originated from North America and has spread to warm regions of the world. The adult is a delicate pale yellow.

Family Blastobasidae

This is a small family of micro-moths comprising a little over 200 species which is found chiefly in tropical and warm climates. The adults are typically dull-coloured and are among the least attractive looking Lepidoptera; they have a characteristic pecten of long hair-like scales at the base of the antennae. The larvae are scavengers, feeding on seeds and decaying vegetable matter; and as a result get carried around with commerce. One species, *Blastobasis lignea*, which originated from Madeira, where the family flourishes in the mild climate, has become established and widespread in the British Isles and has reached Australia. Species of the genus *Holocera* are of particular importance since they have larvae which are predaceous on scale insects and can be used for biological control.

Family Xyloryctidae

This family and the allied one Stenomidae have often been placed together but recent taxonomic studies have produced evidence that they should be separated. It appears that xyloryctids are mainly Old World in distribution and are centred on the Australasian region, while stenomids are New World and are centred on the Neotropical region.

The xyloryctid moths are robust and rather large for Microlepidoptera, having more the appearance of heavy-bodied noctuid moths. Like the noctuids they are night-fliers and come readily to light. The species of the genus *Cryptophasa* are among the largest and many occur in Australia and Papua New Guinea.

Family Momphidae

A large worldwide family comprising over 1000 species the adults of which are small and narrow-winged. The family may be divided into several subfamilies, such as Cosmopteriginae and Chrysopeleiinae. Adults of the Cosmopteriginae are among the most spectacular, having very slender pointed wings, the forewing often being adorned with brilliant metallic gold, silver or orange markings.

Momphid larvae usually mine in leaves, among seeds, or in stems, a few cause galls, and in the tropics a number play an important role in feeding on dry vegetable refuse or act as scavengers for other species. A few momphids rate as pests and among these are species of the genus *Blastodacna* whose larvae damage the young shoots of apple trees. One species which is sometimes a nuisance on stored grain is *Sathrobrota rileyi*, whose larva is normally a scavenger on spent cotton bolls, maize, sugar cane and banana plants. Originating from North America this species has been carried to various parts of the world.

Family Scythrididae

This is a fairly small family of micro-moths of worldwide distribution which are most numerous in Europe, Africa and Asia. The adults are generally uniform in character, with somewhat attenuated wings and dark coloration. Some of the larvae live in long silken galleries spun along the stems of ling (*Calluna*) and heaths (*Erica*) and other herbaceous plants. One of the commonest European species (none is very common) is *Scythris grandipennis*, whose grey-green larva lives in a silken web on gorse (*Ulex*).

Superfamily Tortricoidea
Family Tortricidae (Leaf-roller Moths, Bell Moths, Fruit Moths)

This vast family of micro-moths consists of 4500 described species, and many more await discovery in parts of the tropics. It may be subdivided into several subfamilies, the largest and most important of which are the Olethreutinae, Tortricinae and Cochylinae. In some classifications the latter group is treated as a separate family.

The family Tortricidae is worldwide in distribution and is equally well represented in the tropics and temperate regions. It contains many economically important agricultural and forestry pests. Larvae of the subfamily Tortricinae typically roll and twist leaves together with silk, so as to form shelters in which they live.

One of the commonest and most distinctive tortricids in Europe is the Green Oak Tortrix, *Tortrix viridana*, easily recognizable by its bright green forewings. The moths are found commonly in nearly every oak wood, sometimes in considerable numbers from early summer onwards. The eggs are laid in small groups on the twigs, where they overwinter before the larvae develop and emerge in the spring. They immediately start feeding on the young leaves, eating irregular holes in them and rolling the edges of the leaves, binding them down with silk. They are very active, wriggling violently if disturbed and dropping off the leaf, hanging by a silken thread. When danger has passed they proceed to hoist themselves up to the leaf, by repeatedly gripping the strand of silk between their mandibles and then lifting themselves so that the slack can be taken up by the thoracic legs.

An Australian tortricid, the Light-brown Apple moth, *Epiphyas postvittana*, which is a pest in the apple orchards of the Australian coastal regions, has a larva which is polyphagous and feeds on a variety of trees, shrubs and herbaceous plants. As a result it has been introduced to other parts of the world and in some countries has become established, including the south of England.

The Olethreutinae include one of the greatest and most widespread pests of apple in the world. This is the Codling moth, *Cydia pomonella*, the larva of which can decimate apple crops, and also attacks the fruit of pear, apricot, plum and medlar. Partial control of this species can be effected by promptly gathering up fallen fruit and mulching it, and by searching store boxes for the cocoons.

A closely related species which is also a major pest is the Oriental Fruit moth, *Cydia molesta*. Originally from the Far East, this species has now colonized much of Europe, North America and Australia. The larva will feed on various fruit trees, including apple, apricot and quince, but has a strong preference for peach. Several generations may occur in a year, and larvae of the early ones bore into the twigs causing leaf fall while the later ones enter the fruit. Like that of the Codling Moth, the larva spends the winter spun up in a cocoon in crevices of the bark of its hostplant, and is also to be found in fruit crates.

Also in the same genus is the notorious Pea moth, *C. nigricana*, a European species introduced to North America, the larva of which is a pest of cultivated peas. The adults appear in the spring and lay their eggs on the flowers or developing pods, in which the larva lives concealed from sight. In Europe the fruits of the sweet chestnut and walnut are attacked by the larva of *Cydia splendana*, and the acorns of oak trees by *C. fagiglandana*. The famous Mexican Jumping Bean moth, *C. saltitans*, is a close relative of these species. The 'beans' are the seeds of a Mexican spurge and the moth larva lives inside them. After consuming the contents of a seed the larva lines the cavity with silk. This usually coincides with the time when the pods are ready to burst and throw out the marble-like seeds. The jumping is done by sudden violent contracting and relaxing movements of the larva, causing its body to jerk against the wall of the seed and thus making it 'jump'.

A number of olethreutines are pests of coniferous trees. The Pine-shoot moth, *Rhyacionia buoliana*, attacks Scotch pine in Britain and Europe and in North America. It is a handsome moth, with a wingspan of about 2 cm (0.7 ins), the forewings being orange-brown variegated with silvery white in an irregular pattern. The larva attacks the buds, causing the shoots to shrivel and become deformed. Other common olethreutines on conifers in Europe are the Pine Resin-gall moth, *Petrova resinella* and *Epinotia tedella*. The larva of the first species bores tunnels in the shoots, usually living two years, and its presence is betrayed by the masses of resin that weep from the wound. The larva of the second species attacks fir trees (*Abies*), mining the needle-like leaves and causing them to turn brown and eventually to die.

Three species of Tortricidae are frequent pests of the grapevine. They are the European Grape moth, *Lobesia botrana*, the Grape moth, *Eupoecilia ambiguella*, and *Sparganothis pilleriana. L. botrana* is the most serious and sometimes causes extensive damage in vineyards in the Mediterranean countries. The adults appear in late spring and the females lay their eggs on the flower buds which the larvae envelope with strands of silk and eat.

Superfamily Alucitoidea
Family Alucitidae (Many Plumed Moths)

This is a small worldwide family most numerous in Africa and the Indo-Malaysian region. Alucitid moths are immediately recognizable by their extraordinary wings which are each cleft into six or seven feather-like lobes. The moths are very fragile and are weak fliers. The few European

species are usually found in moist, shady situations, such as the entrances to caves, grottoes or tunnels. The most frequently seen species in Europe is the Twenty-plume moth, *Alucita hexadactyla*, the larva of which feeds in the flower buds of honeysuckle (*Lonicera*). The moth is out from late summer to autumn, after which it goes into hibernation, hiding among ivy or creepers or in dark corners of sheds and outhouses until the following spring.

Family Carposinidae

This small family is found mainly in Australasia and Hawaii. The adults are mostly medium-sized micro-moths, with cryptic white, grey and black coloration and markings, often with roughened scaling on the forewing, enabling them to rest concealed among lichens. *Carposina berberidella*, whose larva lives on the fruit of barberry (*Berberis vulgaris*) occurs in Europe. In Australia, a great many species are found in the sclerophyll forests, and one lives in the bark of the eucalyptus.

Family Copromorphidae

A small family which is centred on Australia and Papua New Guinea, with outliers in the Pacific islands. Some of the adults are moderately large for Microlepidoptera, and have roughened scaling to the forewings. Specimens of this family are rare in most collections and hardly anything is known of the biology of the various species.

Superfamily Pyraloidea
Family Pyralidae (Meal Moths, Wax Moths, Bee Moths, Grass Moths, Knot-horns, Tabbies)

Distributed worldwide, this is perhaps the largest family among the Lepidoptera, with numerous species in all regions. The vast majority of the adults are small to medium-sized, and show an infinite variety of wing shapes, coloration and pattern, but the typical pyralid has rather broad triangular wings often with a translucent appearance, and long spindly legs with long spurs. The family is divided into a number of distinctive subfamilies.

The subfamily Crambinae has a worldwide distribution and includes the Grass moths and Cane Stem-borer moths, which are especially characteristic of the temperate regions. The adults usually have long, narrow wings which they wrap tightly round the body when at rest, which gives them a somewhat cylindrical appearance and makes them difficult to detect when resting on grass stems. A common Grass moth found in Europe and North America is *Crambus*

Right : The Oak Eggar moth, Lasiocampa quercus *(Lasiocampidae), a common European species whose imagos fly in hot sunshine.*
Below : The Eri Silk moth, Samia cynthia *(Saturniidae), originally a native to China, has over the past few decades become established in parts of southern Europe.*

perlella, which is distinctive in having pearly white fore-wings, sometimes streaked with brown. This subfamily includes many species of rice, maize and sugar-cane stem-borers. In Africa and Asia these are of the genus *Chilo* and in Central and South America and the West Indies they are of the closely related genus *Diatraea*.

The subfamily Schoenobiinae also contains a number of economically important stem-borers of rice and wheat especially of the genus *Scirpophaga* in the Indo-Australian region. The adults are typically white or yellowish-white moths with rather pointed wings. Among them is a remark-able aquatic moth known as the Water Veneer, *Acentria nivea,* which occurs in Europe and North America. The larva feeds on various water plants in ponds and lakes and lives and pupates underwater; respiration takes place through its skin. When fully grown it spins a silk cocoon attached to a plant at depths of up to a metre. The cocoon contains air which apparently it obtains from the plant. The male moths are always fully winged, but there are two types of female, one with normal wings which on emerging from the pupa leaves the water and is terrestrial, the other with reduced wings which 'flies' or swims underwater. The terrestrial females and the males can be observed at dusk flying over the water.

The subfamily Nymphulinae includes the China-Mark moths among which are a number of species whose larvae, and sometimes pupae, are also aquatic. A common species in Europe and Asia is the Brown China-Mark, *Nymphula nymphaeata,* which lives on *Potamageton* and various other water plants, preferring ponds and lakes where the water is still. At the beginning of summer the females emerge and after pairing lay their eggs on submerged leaves, into which the larva bores. Later the larva constructs a flat raft-like case made up of leaf fragments, and floats on the surface from one leaf to another.

The subfamily Pyraustinae is very large and includes numerous species awaiting proper classification. It also contains one of the most economically important pyralids. This is the European Corn-borer, *Ostrinia nubilalis.* Originating from the Palaearctic region this species was introduced into North America early in the twentieth

century and rapidly spread to all the maize-growing areas with disastrous results. Besides maize, the larva also attacks tomatoes, sugar-cane and hemp and at high population density and shortage of food it will move onto beans and other low growing plants and even apple and pear trees. On maize, the eggs are laid on the leaves, and the hatched larvae bore into the stem and from there they manage to penetrate the cob itself.

An attractive member of the Pyraustinae is *Palpita unionalis*, which belongs to a large group with white or green wings often lustrous or translucent and with beautiful patterns. *P. unionalis* is one of the plainer species with white, mother-of-pearl wings, with a yellow edge to the front margin of the forewing. Its larva feeds on the foliage of olive and jasmine. A much less attractive-looking moth, with dull brown coloration and an obscure forewing pattern, is the Rush Veneer, *Nomophila noctuella*. Despite its small size—about 30 mm (1 in) wingspan—this moth is a very strong flier and migrates annually from North Africa and Europe to Britain. In North America it is now known as the Celery Webworm moth, *Nomophila nearctica*, while the one occurring in Australia has been appropriately named *N. australica*.

The subfamily Galleriinae includes the Wax moths, Bee moths and Honeycomb moths, most of which have rather sombre yellowish or brownish coloration, sometimes with a touch of green or red. The Wax moth, *Galleria mellonella*, is found in Europe, including Britain, North America and Australia. The larva can prove troublesome to bee-keepers, since it feeds on the wax of the honeycombs. The Lesser Wax moth, *Achroia grisella*, which has also spread throughout the world also feeds on the wax in honeycombs. The Bee moth, *Aphomia sociella*, usually lives in the nests of bumble bees and wasps, but occasionally enters hives. It is not so widespread as the previous species, but occurs throughout Europe and in North America.

The subfamily Phycitinae includes the Knot-horn moths and is found worldwide and is very numerous in species, many of which are of major economic importance. One such is the Indian Meal moth, *Plodia interpunctella*, which is a pest of stored food. The larva feeds on almost any household foodstuff and spoils the food by covering it with tough silken webbing. The species requires warm conditions in which to flourish and will then breed continuously; it cannot normally survive long out-of doors in temperate or cold climates. A number of species of the genus *Ephestia* are also pests of stored food. One of the commonest is the Mediterranean Flour moth, *E. kuehniella*, which despite its name is cosmopolitan and likely to turn up almost anywhere in bakeries, warehouses, shops and domestic food stores generally getting into bread or cakes. Its relative the Cacao Moth, *E. elutella*, prefers dried fruit, nuts, cocoa and chocolate but will go onto grain and cereals, while the Dried Currant Moth, *E. cautella*, prefers raisins and other dried fruits.

Among the numerous phycitid moths there are a few that are at times useful to man. An outstanding example is the Cactus moth, *Cactoblastis cactorum*, a species that was imported from South America into Australia to control and limit the growth of the Prickly-pear cactus (*Opuntia*). This introduced cactus had spread out of control in Queensland and had overrun millions of hectares of grazing land. The importation and release in large numbers of the Cactus moth produced dramatic results. The moth larvae ate into the succulant parts of the cacti, causing them to wilt and eventually rot; thus most of the grazing areas were reclaimed.

Superfamily Pterophoroidea
Family Pterophoridae (Plume Moths)

Species of this family are found in all parts of the world but because of their strange unmothlike appearance and retiring habits they are seldom noticed. The adults are recognizable by their deeply fissured wings and long, spidery legs, the forewings being cleft into two, or more rarely three or four plumes, and the hindwings into three plumes, with the

Above: The caterpillar of the rare Spanish Moon moth, Graellsia isabellae (Saturniidae).
Right: The imago of the Spanish Moon moth. This species is found only in Spain and the south of France.

exception of the few species comprising the subfamily Agdistinae where the wings are long and narrow but are not cleft.

Most pterophorids fly at sunset and are feeble fliers with little muscle-power; but their extreme lightness of structure enables them to float in the slightest breeze and to be carried considerable distances. A number of species are found on mountains, the adults flying in the brief periods of calm; in South America a species has been caught at over 4000 m (13,122 ft). In Europe one of the commonest species in gardens is the White Plume moth, *Pterophorus pentadactyla*, which has white wings; in the summer it often comes to lighted windows. The larva feeds on bindweed (*Convolvulus*), and the moth is therefore a useful species to have in the garden. Pterophorid larvae are typically rather short and hairy, and usually feed exposed on flowers or leaves, but sometimes penetrate the stems and seed capsules.

Superfamily Bombycoidea
Family Lasiocampidae (Lappets, Lackeys, Eggars, Tent Caterpillar Moths)

These rather robust, large-bodied and often cumbersome-looking moths are found in most parts of the world but are apparently absent from New Zealand. The larvae have a dense coat of secondary setae and sometimes bear urticating hairs or bristles; some are flattened and have a series of lateral tufts.

The Lappet moth, *Gastropacha quercifolia*, which occurs from Europe, including Britain, to China and Japan, at rest resembles a bunch of brown leaves. Its large, hairy, grey caterpillar feeds on willow, apple and a variety of other trees, and has been known to eat the poisonous leaves of the laurel. An important European species is the Lackey moth, *Malacosoma neustria*, which sometimes effects orchards, forests and hedgerows. The female lays her eggs in July and August, depositing them in broad bands round the twigs of the hostplant. The eggs remain dormant through the winter

and hatch the following spring. The caterpillars are covered with red, blue, yellow, white and black hairs and live in communal nests or tents of silk spun among the thinner twigs or branches.

The genus *Lasiocampa* is confined mainly to Europe, Africa and Asia. Two common European species are the Grass Eggar, *L. trifolii,* and the Oak Eggar, *L. quercus.* The first feeds mainly on leguminous plants, particularly clover, while the second lives on various shrubs and herbaceous plants, but prefers ling, heather and bramble. In both of these species the male has russet-brown wings with a broad transverse band of yellow of the forewing, and the female is yellow-ochre with a pale transverse band. The moths usually emerge from the pupa about midday during the summer months, and the males are soon on the wing in search of females. Males of the Oak Eggar in particular are likely to be seen flying wildly over rough heathery country in early summer, with an erratic zigzag flight. After pairing the females take wing and start dropping their eggs at random as they fly.

Family Bombycidae

Only a few species belong to this small Asiatic and oriental family of moths. The most important is the Cultivated Silk moth or Silkworm, *Bombyx mori.* The adult is of a rather unprepossessing appearance for such an important species, with moderately plumose antennae in the male, and rather broad, slightly falcate forewings which are dull white except for the brownish veins.

The caterpillar varies in colour, and is swollen near the head; it feeds on black or white mulberry (*Morus*) and will accept garden lettuce as a temporary alternative. It spins a silken cocoon which centuries of selective breeding has gradually improved for commercial purposes so that it yields more than a kilometre of high quality unbroken thread. During the process of domestication, however, the adults have lost their power of flight, and the species is no longer found in the wild state. In the genus *Bombyx*, as in

Below : A typical geometrid moth with faintly coloured wings which are large in comparison to its small body.
Right : The Death's-head Hawk-moth, Acherontia atropos (Sphingidae), which owes its vernacular name to the skull-like design on the dorsal part of its thorax.

many other bombycoids, the frenulum and retinaculum have been lost, and the wing coupling is amplexiform.

Family Saturniidae (Silk Moths, Emperor Moths)

This is a family of giant silkmoths and emperor moths, mainly tropical in distribution but extending into the temperate regions. The largest moth in the world, in wing area and bulk, is reputed to be the Hercules moth, *Coscinocera hercules*, which lives in the rain forests of tropical Australia and Papua New Guinea. A female of this species taken in Queensland in 1948 was reported to have a wingspan of 36 cm (14 ins). Rivalling it in size are the females of the Atlas moth, *Attacus atlas*, from the foothills of the Himalayas. Saturniids are typically brightly coloured, with a conspicuous eye spot or a little transparent 'window' on each wing, and the males have plumose or feathery antennae. The larvae may be huge and bulky, measuring 10 cm (4 ins) or more in length, with a girth exceeding 2 cm (0.7 ins). Many are bright green, with prominent coloured tubercles bearing bristles or spines which if touched can cause severe skin irritation. The larvae spin huge silken cocoons, sometimes mixed with larval hairs. Those of the species in the genus *Antheraea* have been cultivated for hundreds of years in China, India and Japan for commercial silk production. Shantung silk or Chinese tussore silk is still exported from China, and the production of tasar silk is a flourishing industry in India. In North America the commonest saturniid is probably the Polyphemus moth, *Antheraea polyphemus*. It is often bred in captivity, the larva feeding on apple and hawthorn leaves. Fully grown caterpillars are green, with red tubercles.

One of the best known species of this genus in Australia is the Emperor Gum moth, *Antheraea eucalypti*, whose large green larvae are often found on eucalyptus. Like those of some other saturniids, the pupa may remain dormant in its cocoon for several years before the moth emerges.

The Eri Silk moth, *Samia cynthia*, was recently introduced into southern Europe from China and has become established in the wild. The caterpillars feed on the tree of heaven (*Ailanthus*) and the castor plant (*Ricinus*).

The Giant Emperor moth or Great Peacock moth, *Saturnia pyri*, is the largest native European moth or butterfly, but it is not found in Britain; the females attain a wingspan of 16 cm (6 ins). The larvae are vividly coloured, with alternating blue and yellow stripes and are adorned with large blue tubercles. They feed on the foliage of apple and various other trees. The similar but smaller Emperor Moth, *S. pavonia*, does occur in Britain, ranging throughout Europe to Siberia and Asia. The males are day-fliers and are most often to be seen over heathland; the females usually only fly at night. The caterpillar feeds on ling (*Calluna*), bramble (*Rubus*) and bilberry (*Vaccinium*).

One of the most distinctively patterned saturniids found in alpine beech woods in Europe, is the Tau Emperor, *Aglia tau*. The adult is brightly coloured and may attain a wingspan of 7 cm (2.75 ins). The wings are typically yellow-ochre with a metallic blue patch on each, in the centre of which is a white mark resembling the Greek letter *tau*. Specimens vary in colour, and some may be brown or black and with small or large eye spots.

A rarity in Europe is the Spanish Moon moth, *Graellsia isabellae*, a large elegantly marked moth with yellow-green wings. In spite of its exotic appearance this species is strictly European, as far as is known, and is only found in a few localities in central Spain and the French Alps.

Family Anthelidae

A speciality of the Australian and Papuan faunas; the moths of this family of about 100 species are reminiscent of the Lasiocampidae in general coloration and appearance. Their larvae are also often covered with short hairs which are liable to break off and cause skin irritation.

Family Brahmaeidae

This rather select family comprises only a few species which are mostly scattered through the Ethiopian region and Asia. The adults are generally large and moderately robust and

rather saturniid-like in appearance. Only one species, *Acanthobrahmaea europaea*, which is of moderate size is known in Europe.

Superfamily Geometroidea
Family Geometridae (Geometrids, Loopers, Inch Worms, Emeralds, Waves, Carpets)

Found worldwide, this is one of the largest families of Lepidoptera, numbering some 12,000 species which are divided into several subfamilies. The adults are generally small to medium-sized, with wings that are flatly expansive in relation to the slender body and often marked with a complicated transverse linear pattern. Most geometrid moths fly at dusk and at night. The larvae have a characteristic appearance: apart from the thoracic legs, which are invariably present, they never have more than two pairs of abdominal prolegs, usually on the sixth and tenth segments —in most other families of Lepidoptera the caterpillars have four pairs of prolegs. Consequently, geometrid larvae have a curious mode of progression, first bringing the tip of the abdomen up to the thorax in a loop, clasping the substrate with the hind prolegs and then moving the thorax forward, curving and then extending their back in a manner reminiscent of leeches. As they advance, they seem to be measuring

Above: The Convolvulus Hawkmoth Agrius convolvuli (Sphingidae), has a very long proboscis which it uses to suck nectar from convolvulus, petunias and other tubular flowers.

the ground they cover as one does with a pair of compasses; this has earned them the name geometrids. The larvae of many species are remarkable for their mimicry: some look like little twigs down to the finest detail—even to the extent of bearing little bud-like protuberances on their bodies. Furthermore, some of these twig-like larvae can enhance their resemblance to inanimate objects by remaining motionless and rigid for hours on end thus protecting them from any predators.

Subfamily Geometrinae

These are mostly beautiful green-coloured moths, the colour sometimes verging towards yellow. One of the largest is the Large Emerald, *Geometra papilionaria*, which occurs throughout Europe to Asia and Japan and has a wingspan of 5 cm (2 ins). The larva lives on the foliage of birch and various other trees and shrubs, and shows a high degree of mimicry: when hibernating it is brown, but when spring comes and the leaves appear it turns green.

Subfamily Larentiinae

This includes a number of species with their wings decorated with a characteristic design of transverse wavy lines, often meeting in a dark patch at the centre or base of the wing. A well-known species is the Winter moth, *Operophtera brumata*, a defoliator of apple, oak and other garden and woodland trees in Europe and North America. The male is fully winged, with a wingspan of over 2 cm (0.7 ins), while the female is brachypterous. The moths emerge from the pupa in autumn and winter, and the females at once begin to climb laboriously up the trunk of a tree to mate and lay their eggs on the branches or at the base of the buds. In orchards, a popular method of trapping them is to encircle the trunks with a grease-band or some other sticky substance.

Subfamily Ennominae

This large, heterogeneous and widely distributed group includes the boldly marked white with black and orange-yellow, Magpie moth, *Abraxas grossulariata*, which is familiar to most gardeners in Europe who grow currants and gooseberries. The caterpillar is creamy white, marked on the back with black patches, and with lines of dots on the sides. The beautiful, pale yellow Swallow-tailed moth, *Ourapteryx sambucaria*, is one of the largest European geometrids, having a wingspan of 5 cm (2 ins), and is also found in gardens.

An important member of this subfamily is the Peppered moth, *Biston betularia*, which has become internationally

Above: The Lime Hawk-moth, Mimas tiliae *(Sphingidae), one of the commonest of the European Hawk-moths. The normal green and brown wing pattern is very variable and is sometimes tinged with red.*

famous as a result of scientific studies in England and Europe on the phenomenon called Industrial Melanism. Another member of this subfamily is the Bordered White Beauty, *Bupalus piniaria*, a geometrid that is destructive to pines in central Europe. The caterpillars sometimes cause extensive damage in plantations, completely defoliating individual trees.

The adults show marked sexual dimorphism, the male having brown wings with a yellowish pattern, while the wings of the female are a more or less uniform orange colour, contrasting strongly to those of the male.

Family Uraniidae

This small family consists of about 100 species, confined mainly to the tropics. Most of the moths are medium to large, and many are diurnal or crepuscular in habit, except members of the subfamily Microniinae which are flimsy and white and are nocturnal. The eight or so species of the tropical American genus *Urania* show magnificent shades of blue, green, red and gold and have short tails to the hindwings, which often leads to them being mistaken for papilionid butterflies. The most spectacular species is without doubt the brilliantly iridescent *Chrysiridia ripheus* a native of Madagascar. The background colour of the wings of this species is jet black and is traversed by green and blue stripes. The hindwings carry several little tails and are speckled with purple, while the fringes of the wings have white highlights. Changes in the angle of incidence of the rays of light falling on it result in a magnificent display of colour.

Family Drepanidae (Hook-tip Moths)

A small family consisting of about 800 species, chiefly Old World. The adults bear a certain resemblance to the Geometridae, but can generally be readily distinguished by their characteristic hook-tipped forewings. Two common European species are the Barred Hook-tip, *Drepana cultraria*, which lives mainly on beech, and the Oak Hook-tip, *D. binaria*, which lives on oak. A curious little moth belonging to this family is the Chinese Character, *Cilix glaucata*, which is found commonly in Europe and the Mediterranean region. It is one of several species which when at rest resemble bird-droppings. It is also known as the Goose-egg moth, presumably because of the grey-brown, egg-like patch at the middle of the forewing.

Superfamily Sphingoidea
Family Sphingidae (Hawk-moths)

This worldwide family of nearly 1000 species includes some of the best-known and most popular moths. The larger species have a powerful streamlined form which makes them fast fliers, capable of speeds of 50 kph (31 mph) or more. The family is divided into several subfamilies which include diurnal as well as nocturnal species. Some of the adults are capable of hovering, poised before a flower in the manner of a humming-bird to feed on the nectar with their long proboscis. The sphingid tongue is sometimes very long indeed, up to 25 cm (10 ins) in the American genus *Cocytius*, which contains the Giant Sphinx, *C. antaeus*; or it may be short and heavily sclerotized, or atrophied and not

Left : The distinguishing feature of the Eyed Hawk-moth, Smerinthus ocellata (Sphingidae) is the large blue eye-spot with a black border against the yellow and pink background of the base of the hindwing. Below : The Hummingbird Hawk-moth Macroglossum stellatarum (Sphingidae), a very fast-flying diurnal hawk-moth found throughout the British Isles, in Europe, India, China and Japan.

Left : The Elephant Hawk-moth, Deilephila elpenor *(Sphingidae), and its relative the Small Elephant Hawk-moth,* D. porcellus, *occur in Britain and Europe. They spend their pre-imaginal stages on bedstraws (*Galium*) and other low-growing plants.*

functional. The caterpillars are distinctive in being rough-skinned and nearly hairless and armed with an often savage-looking horn at the rear end. Although this horn is some-times long and sharply pointed it is quite harmless and not used as a weapon. Sphingid larvae have been compared to the Egyptian Sphinx on account of their habit of raising the front part of the body when disturbed or at rest, and remaining motionless in this posture for long periods of time.

Subfamily Sphinginae

One of the largest European members of this subfamily is the Death's Head Hawk-moth, *Acherontia atropos*, so named because of the yellowish markings on the back of its thorax resembling a skull or death's head. The adult may attain a wingspan of 12 cm (4.5 ins), and has rather strange dietary habits : it lives on honey, which it takes from within the nest or hive, using its short, pointed and sclerotized proboscis to pierce the lids of the honey-cells. When forcing its way into the hive or nest it can produce a high-pitched rhythmic squeak or hiss by forcing air through the proboscis. This sound is supposed to calm the bees.

The huge caterpillar feeds on the foliage of various solanaceous plants, including potatoes and deadly night-shade. It is armed with a large tail-horn and is of fearsome

appearance but is harmless to man. The Death's Head seems unable to survive from one year to the next in Europe, but is replaced each spring by recruits from Africa which fly across the Mediterranean to southern Europe and over the Alps to reach Britain and other northern countries.

Probably the commonest and most widespread sphingid is the Convolvulus Hawk-moth, *Agrius convolvulvi*, which has a wingspan of up to 11 cm (4 ins). Its wings are a varie-gated grey, while the abdomen has stripes of pink, black and white. The adult is a long-distance flier, and ranges to the Pacific islands and Australia. It is also remarkable for its extremely long tongue which is developed for sucking nectar from tubular flowers, such as those of convolvulus, petunias and tobacco. An almost identical species belonging to the same genus occurs in America and is known as the Sweet Potato Hornworm, *A. cingulatus*. The Pine Hawk-moth, *Hyloicus pinastri*, damages conifers and is much feared in central Europe. It is smaller than *A. convolvuli*, with greyish brown wings and a black and white pattern on the sides of the abdomen. The Lime Hawk-moth, *Mimas tiliae*, is often a common species in towns and cities in Europe wherever there is a plentiful supply of mature lime trees. The caterpillar is lime-green, with seven oblique yellow stripes on the sides, each edged above with reddish purple and has a blue curved horn. The adult has rather dull green and reddish brown coloration and when at rest resembles crumpled leaves.

A common garden species in Europe is the Eyed Hawk-moth, *Smerinthus ocellata*, which has variegated grey forewings, and hindwings that are pink at the base and bear a large, blue eye-spot edged with black. The caterpillar is green, with oblique white lateral stripes and a light blue horn; it lives on the foliage of apple and willow trees. Even more common and widespread is the Privet Hawk-moth, *Sphinx ligustri*, which ranges from Europe to China and Japan. Its purple and white striped, pale green caterpillar feeds on the leaves of privet hedges. Like those of the Death's Head, Lime Eyed and other hawk-moths, when fully grown it leaves the foodplant and burrows into the soil or among surface leaf-litter to pupate, either in an earthen cell or a flimsy silken cocoon.

Subfamily Macroglossinae

This subfamily includes the Bee Hawk-moths of the genus *Hemaris*, with representatives in Europe, the Mediterranean region and North America. These are beautiful medium-sized hawk-moths and remarkable for the fact that soon after emergence their wings lose most of their scales and become almost transparent. The moths resemble bumble-bees, especially when in flight. The related genus *Macroglossum* includes the Humming-bird Hawk-moths and is more widely distributed with species in Europe through

Asia and the Pacific islands to Australia. It includes the common European species *M. stellatarum* which is known as the Humming-bird Hawk-moth. This species has a wingspan of 5 cm (2 ins) and is migratory and is capable of flying at great speed. The proboscis of this species is very long, and the abdomen is blunt-tipped and bristled.

Another beautiful member of this subfamily is the Oleander Hawk-moths, *Daphnis nerii*. This large moth, which has a wingspan of about 12 cm (5.5 ins), is clad in a sumptuous livery of green, white and pink. It inhabits Africa and the Mediterranean region, but occasionally migrates to northern Europe. The caterpillar feeds on oleander and lesser periwinkle (*Vinca minor*). Among the many other attractive member of this group are the Spurge Hawk-moth, *Hyles euphorbiae*, found in Europe and North America, and the Striped Hawk-moth, *H. lineata*. The latter ranges from Europe to Australia and is represented in America by the subspecies *H. lineata livornica*. Its caterpillar sometimes harms grape vines, but it usually keeps to bedstraws (*Galium*) and Virginia creeper. The Elephant Hawk-moth, *Deilephila elpenor*, a moth with delicate yellow, pink and sooty grey coloration, and its junior the Small Elephant Hawk-moth, *D. porcellus*, which is similar in appearance but smaller, are both widespread in Europe and also feed on bedstraws. The caterpillar of *D. elpenor* tapers towards the head in a distinctly trunk-like way, and has two conspicuous eye-like markings on the thoracic segments.

Superfamily Notodontoidea
Family Notodontidae (Prominents, Lobsters, Puss Moths)

This is a large worldwide family of medium to large moths, numbering perhaps 3000 described species. The adults often have characteristic tufts of raised scales on the hind margin of the forewings and on the back of the abdomen, and the males usually have strongly pectinate or dentate antennae. In most species a tympanal organ is present on the thorax. The larvae often bear tubercles or spines and have the prolegs of the tenth abdominal segment transformed into slender processes which may be very long. This is very clearly seen in the common European Puss moth, *Cerura vinula*, which lives on poplars and willows. In this species, the prolegs or claspers of the tenth segment are developed into a pair of filaments, each of which can extend a long, wine-red tentacle if the caterpillar is disturbed.

An interesting member of the Notodontidae is the European Lobster moth, *Stauropus fagi*, the larva of which lives on beech. The caterpillars are remarkable for their lobster-like appearance, having extremely long, thin meso-thoracic and metathoracic legs, with the abdomen bearing a number of spiny projections and its rear end greatly dilated and bearing two slender appendages. When disturbed it raises the swollen rear end over its back and vibrates the long thoracic legs. Another common European species is the Buff-tip, *Phalera bucephala*, which is usually found wherever lime trees occur. When at rest the silver and pale yellow contrasting coloration of the forewings makes the moth resemble the freshly broken end of a twig or branch.

Family Thaumetopoeidae (Processionary Moths)

This family contains relatively few species but is represented in most regions. Their vernacular name, of processionary

Below: The subfamily Lithosiinae (Arctiidae), includes moths that are uniform and generally yellowish in appearance. This is a species of the genus Lithosia, *commonly known as Footmen moths.*

moth, derives from the gregarious habit of the caterpillars in moving about one behind the other. This behaviour has intrigued many naturalists, and attempts have been made to work out the mechanism behind it. What is clear so far is that the caterpillars are excited by the hairs on the tip of the abdomen of the one in front, and that every caterpillar secretes a silk thread which serves as a guide for those behind. Another feature of these caterpillars is the tuft of irritant hairs on the mid-dorsal area of each abdominal segment. Two species occurring in southern Europe and North Africa are the Oak Processionary moth, *Thaumetopoea processionea,* which lives on oak trees, and the Pine Processionary moth, *T. pityocampa,* which lives on pine trees.

The larvae of both species live in communal webs or nests which are sometimes very large and accommodate several hundred individuals. The larvae emerge from the nest at night and assemble in a wedge-shaped formation with a single caterpillar at the head, and move off in search of food.

Superfamily Noctuoidea
Family Noctuidae (Noctuids, Underwings, Armyworms, Cutworms, Owlets)

An enormous worldwide family, numbering more than 20,000 species which are classified into numerous sub-families, only some of the principal ones being mentioned here. The adults are generally nocturnal and with few exceptions are small to medium-sized moths. One of the exceptions is the Giant Owl moth, *Thysania agrippina,* of tropical America, which has a wingspan of 30 cm (12 ins) and is one of the largest moths in the world. Most noctuids have rather dull-coloured coloration which seems to blend with the background of their surroundings, but there are exceptions and some have bright, metallic coloration, but these are usually day-fliers. A large number have dull-coloured forewings but brightly coloured red, yellow or blue hindwings which is usually flash coloration for confusing a pursuing predator.

Below : A magnificent specimen of the Jersey Tiger moth, Euplagia quadripunctaria *(Arctiidae), bearing black patches on its fiery-red hindwings. This species is distributed throughout southern Europe but becomes scarcer northwards and reaches only the south-west of the British Isles.*

Subfamily Noctuinae

Many species of this large subfamily are agricultural pests, and include the armyworms and cutworms which are among the most notorious pests of crops. One of the most destructive is the Dark Sword Grass or Black Cutworm, *Agrotis ipsilon*, a cosmopolitan insect and highly injurious to such crops as cereals, cotton, hemp, beans and beets. Most of the agrotids prefer rather moist temperate conditions, and in Australia one species has adapted to an unusual life-style to overcome the arid conditions. This is the Bogong moth, *Agrotis infusa*, which is a pest of pastures and winter crops. To avoid the hot dry summer the adults migrate to the mountains and cluster together in caves and rock crevices where they aestivate until the cooler weather returns.

One of the commonest European and North American noctuids is the migratory Setaceous Hebrew Character, *Xestia c-nigrum*.

Subfamily Hadeninae

The moths of this subfamily may be distinguished from other noctuids by the presence of hairs on the surface of the compound eyes; the eyes of other noctuids are glabrous. A common European species is the Cabbage moth, *Mamestra brassicae*, which is a minor pest of brassicas.

Subfamily Acronictinae

The species included in this subfamily have caterpillars that are markedly different from those of the majority of other noctuids, being brightly coloured and hairy. The Grey Dagger, *Acronicta psi*, of Europe and North Africa, which derives its name from the grey coloration of the forewings and a black dagger-like marking on them, has a very striking caterpillar. This has a conspicuous clear yellow stripe along its back, with black-edged red spots either side of it, and a tall black hump just behind the thorax; it feeds on the foliage of apple and other orchard trees.

Subfamily Heliothidinae

A small but economically important group. Two of the most destructive pest species are the Old World Bollworm, *Heliothis armigera*, which occurs in Europe, Afric, Asia and Australia, and the American Bollworm or Corn Earworm, *H. zea*, which is widely distributed in America, including

Below : The Garden Tiger, Arctia caja *(Arctiidae).*
Right : The Cream-spot Tiger, A. villica. *These are among the largest arctiids found in Europe. The forewings are striped brown and white in the former, and black with white patches in the latter.*

Left : This plusiine moth which is known as the Gold Spot, Plusia festucae (Noctuidae), has rather disruptive markings, making it difficult to detect when it is resting on a tree trunk or dry branch.

Left : The Setaceous Hebrew Character moth Xestia c-nigrum *(Noctuidae), is widely distributed; it is found in Europe, India, Japan and North America. Its caterpillars attack various garden plants.*

Hawaii. The larvae of both species attack cotton bolls and corn ears as well as other field crops.

Subfamily Eublemminae

These moths are small and some of them show a singular type of behaviour as larvae. That of *Coccidiphaga scitula*, a Mediterranean and oriental species, is predaceous on scale insects. The female moths lay their eggs either on the bodies of their host or in their vicinity. The larvae then gradually consume the colony, and when fully fed spin some of the empty follicles together to form a cocoon in which to pupate.

Subfamily Plusiinae

This homogeneous group has a number of wide-ranging and migratory species. Many of the adults are characterized by the presence of an area of metallic colouring on the fore-wings, as is found in the common Palaearctic Burnished Brass, *Diachrysia chrysitis*. A less striking example is the Silver-y moth, *Plusia gamma*, an almost cosmopolitan species, whose larva is polyphagous on herbaceous plants. The adult is a diurnal and nocturnal flier and regularly migrates over great distances. Its larva, like those of other

plusias, has the first and second pairs of abdominal prolegs more or less aborted, and is known as a semilooper.

Subfamily Catocalinae

This group centres around the large genus *Catocala* or Underwing moths which comprise about 200 species, more than half of them occurring in North America. Their hind-wings are generally bright red or blue, with black borders or bands. The larvae feed on the foliage of willows, oaks, alders and other broad-leaved trees, and like those of the Plusiinae have the first and second pairs of abdominal prolegs aborted.

Family Lymantriidae (Tussock Moths)

This family is worldwide and comprises about 2000 mostly medium-sized species, many of which show marked sexual dimorphism. The males have bipectinate antennae and are fully winged, while the females have short, finely dentate antennae and are often larger but sometimes apterous or brachypterous. The proboscis is non-functional, and the body and legs are covered with dense hairs. The female uses the hairs on the tip of her abdomen to cover and protect the egg batches. The caterpillars are very hairy and often

Above: The Burnished Brass, Diachrysia chrysitis *(Noctuidae), a plusiine moth whose caterpillars are sometimes destructive to herbaceous plants.*

Overleaf: Three tropical papilionid butterflies found in Papua New Guinea, gather on damp ground in order to absorb moisture and minerals from the soil.

brightly coloured; in some species the hairs are irritant. Many have rows of tufts or tussocks on their backs, like the bristles of a toothbrush.

A well-known lymantriid, and one of considerable economic importance, is the Gypsy moth, *Lymantria dispar*. The male, which may attain a wingspan of 40 mm (1.5 ins), is brown; the female is considerably larger and her wings are variegated whitish yellow. The species is of Palaearctic origin, but became established in North America in the mid-nineteenth century after escaping from a laboratory. There it quickly spread and became a major pest in forests, the larvae causing severe defoliation. A related species, the Black Arches, *L. monacha*, is a pretty white moth with small, black lines and patches on its wings. In Europe it occasionally wreaks havoc in coniferous forests and is a minor pest in orchards.

Many lymantriids are almost pure white and lack distinctive markings. The genus *Leucoma* contains about 50 such species, one of which is the Satin Moth, *L. salicis*, originating from Europe. In North America it has become a minor pest of willow and poplar.

Family Ctenuchidae

A large, mainly tropical family of about 3000 species of mostly small to medium-sized, narrow-winged moths, often

brightly coloured and with iridescent scales. They are mainly day-flying and some have the appearance of colourful wasps and show a high degree of mimicry. The species are most numerous in South America, and among the few found in temperate regions is the European *Syntomis phegea*.

Family Arctiidae (Tigers, Ermines, Footmen)

This is a large family of about 10,000 species of mostly medium to large-sized moths, often with boldly marked wings. They are distributed worldwide but are most numerous in South America. Some of the adults have a habit of emitting droplets of haemolymph mixed with air; this has a nauseating smell, and they produce it when touched. The process has been given the name haemaphorrhoea.

Arctiid larvae are often known as woolly bears and are generally very hairy.

Subfamily Arctiinae

These are the largest and among the most spectacular of the arctiids. They include the Garden Tiger moth, *Arctia caja*, a common European species whose hairy, dark brown larva feeds on weeds and garden plants and is the Woolly

Bear, is a robust moth with a wingspan of about 7 cm (2.75 ins). The forewings are creamy white with irregular thick chocolate brown markings, and the hindwings are red, marked with deep-blue-centred black spots which are often ringed with yellow; the abdomen is also red, marked with black. The moths are nocturnal and seldom fly until after midnight, emerging during the summer months. Two other European species of this genus are the Cream-spot Tiger, *A. villica* and Jersey Tiger, *Euplagia quadripunctaria.* The latter is common in continental Europe but is rare in Britain and found only in south Devon and the Channel Islands.

A common European arctiid often found in gardens is the White Ermine, *Spilosoma lubricipeda.* Like other Ermine moths, it is much less spectacular in appearance than the Tigers, being smaller and with essentially white wings only lightly spotted with black. The larva is brown and hairy but is distinguished by a reddish stripe along the middle of its back. Another common European species, often found in rough fields and pastures where its foodplant ragwort (*Senecio*) abounds, is the Cinnabar moth, *Tyria jacobaeae.* This is one of the unpalatable species, both in the larval and adult stages. The larva is perhaps more spectacular than the red and black adult, being gaudily ringed with alternating black and yellow bands. It is often to be found in numbers feeding openly on the ragwort plants, which it can afford to do because of its highly noxious qualities which make it

distasteful to European birds. Surprisingly, however, New Zealand birds found it palatable—perhaps lacking the training and experience of their European counterparts—and proceeded to devour the larvae which were taken over to New Zealand and were released in large quantities in a massive effort to control the alien ragwort which had colonized New Zealand.

Some of the more showy species are found in the genus *Utetheisa,* and includes the Crimson Speckled Footman, *U. pulchella,* which has a wingspan of about 4 cm (1.5 ins), the forewings being white conspicuously speckled with crimson and black. It is a migratory species and one of the most widespread in the Old World, ranging from Europe to Australia. Most archids have large, well-developed tymbal organs in the thorax.

Subfamily Lithosiinae (Footmen)

Generally much less spectacular in appearance than the Tigers, the moths of this subfamily characteristically have rather long, narrow wings with drab coloration. Many of them fold the wings flatly along the body and then have a very smooth and stiff appearance which probably is why they are called Footmen. The larvae are generally short-haired, and many live on lichens.

Those of the genus *Setina,* which includes the Dew moth,

S. *irrorella,* can produce a soft, ticking sound when flying which can be heard a few metres away.

Family Nolidae

Comprising several hundred species, the moths of this group, which is sometimes treated as a subfamily of the Arctiidae, are of rather small size, having a wingspan seldom exceeding 10 mm (0.4 ins). They are generally white with grey or black markings, and are often mistaken for Microlepidoptera. When at rest they usually sit with the head downwards. The caterpillar has only four pairs of prolegs, and lacks the pair on the third abdominal segment.

Superfamily Hesperioidea
Family Hesperiidae (Skippers)

A large, homogeneous worldwide group generally classified as butterflies rather than moths, although they do not have the typical 'clubbed' antennae. The adults are mostly of moderate size, with a stout body and a large head relative to the wings. The antennae are characteristically tapered at the apex and bent backwards to form a hook-like tip. The adults are very active day-fliers and have a rather characteristic skipping flight. The caterpillars are almost hairless and many feed on grasses, living either in a shelter made of leaves spun together, or concealed in a web of silk. Pupation always takes place within this covering. Two subfamilies are known in Europe: Pyrginae and Hesperiinae. One of the commonest in the former group is the Grizzled Skipper, *Pyrgus malvae,* which is found in meadows and bogs from sea-level to high up in the mountains. The pale green caterpillar is often found on strawberry plants, pulling the edges of a leaf together and fastening them down with silk. A common species representative of the second subfamily is the Essex or European Skipper, *Thymelicus lineola.* This species also lives in meadows from sea-level to high up in the mountains.

The Skippers pose some intriguing problems of classification. The Giant or Yucca skippers of America belonging to the genus *Megathymus* have rounded tips to the antennae. They are sometimes placed in a separate family, or in the mainly tropical American family Castniidae. It seems unlikely that they belong to the true Skippers especially as their life histories and larval habits are quite different. The caterpillars are borers in Yucca and Agave plants, living in the stems and roots. The Regent Skipper, *Euschemon rafflesia,* of Australia, would seem to be half moth and half butterfly since only the male possesses a moth-like wing coupling apparatus consisting of a frenulum and retinaculum.

Superfamily Papilionoidea
Family Papilionidae (Swallowtails, Swordtails, Birdwings, Apollos)

This worldwide family contains some of the largest and most beautiful butterflies in the world, attractive not only for their colour and size but also for their graceful flight. The caterpillars are generally hairless, often brightly coloured, and carry spines or fleshy tubercles. Their prothorax bears an eversible forked process, the osmeterium, which emits a pungent scent; this appears to be a defensive mechanism against predators. The pupae or chrysalids are short and often bear two pointed projections on the head.

Subfamily Papilioninae

This large group of butterflies is most prevalent in the tropics and subtropics and includes the swallowtails and birdwings. The genus *Papilio* contains about 200 species, the largest of which is the African Giant Swallowtail, *P. antimachus,* which has a wingspan of up to 25 cm (10 ins).

Far left : A newly-emerged Large White butterfly, Pieris brassicae *(Pieridae).*
Left : Another pierid butterfly of the genus Colias, *distinguished by its broad yellow wings which have a conspicuous dark border.*

It inhabits the tropical rain-forests of Africa and is an enormous butterfly, nearly as large as the Malaysian birdwings. A more modest creature is the European and North American Swallowtail, *Papilio machaon*, with a wingspan of barely 10 cm (4 ins) and yellow and black coloration, enhanced with blue shading on the hindwing and a little blood-red patch at the inner angle. It is generally most common in fenland. The caterpillar is green with black rings speckled with yellow, and feeds on the leaves of fennel and wild carrot. Another handsome species is the Scarce Swallowtail, *Iphiclides podalirius*, which occurs almost throughout the warmer parts of the Palaearctic region and its often seen in gardens where fruit trees grow. Its caterpillar is green with a yellow dorsal streak, and feeds on the foliage of various roses.

Most spectacular in size and coloration are the birdwing butterflies of the tropical forests of the Malaysian, Papuan and northern Australian regions. One of the largest is Queen Victoria's Birdwing, *Ornithoptera victoriae* of the Solomon Islands, the females of which are very much larger and less colourful than the males and are reputed to have wingspans approaching 30 cm (12 ins), but this is probably exceptional and the average wingspan is around 20 cm (8 ins).

Another interesting species is Rajah Brooke's Birdwing, *Trogonoptera brookiana*, which occurs in Malaya and Sumatra. The butterflies inhabit the upper canopy of the rain forest, but come down to feed on flowers and obtain nutrients in muddy glades along streams and rivers. The caterpillars of most birdwings feed on the foliage of poisonous Aristolochiaceae plants and are themselves poisonous to predators.

Subfamily Zerynthiinae

Comprising a dozen or so Palaearctic and oriental species. The Spanish Festoon, *Zerynthia rumina*, frequents rocky slopes on hills and mountains in south-west Europe and North Africa. Its close relative the Southern Festoon, *Z. polyxena*, occurs in similar habitats but ranges eastwards to Asia Minor. The caterpillars of both species feed on *Aristolochia*.

Subfamily Parnassiinae

Known as Apollo butterflies, this group is centred around the genus *Parnassius* which contains about 30 species distributed chiefly in the mountainous regions of Europe, Asia, Japan and North America. The adults have parchment-like wings attractively patterned with red, yellow or black patches. Many of the species are highly variable and show considerable subspeciation, perhaps due to their isolation in

Right : The Brimstone butterfly, Gonepteryx rhamni *(Pieridae), whose forewings have distinctive sickle-shaped tips.*
Below : The Black-veined White, Aporia crataegi *(Pieridae), which is easily recognized by its white, almost transparent wings with their contrasting black veins.*

different mountain retreats. A number of the species are
truly alpine, and *P. acco* has been taken on Mt Everest at
over 5000 m (16,404 ft). The Small Apollo, *P. phoebus,* is
found only at high altitudes in the European Alps and also
in North America as far south as the mountains of Mexico,
and in the Soviet Union. Its caterpillar feeds on stonecrops
(*Sedum*) and saxifrages (*Saxifraga*). The Apollo, *P. parnassius,* is subalpine and occurs at lower elevations in the
mountain regions of Europe, except Britain, eastwards to
central Asia. Its caterpillar is black and hairy, with many
red spots, and also feeds on stonecrops. The Clouded Apollo,
P. mnemosyne, has a distribution similar to that of the Apollo,
but is a lowland species in the north where it frequents
damp meadows. It is readily distinguished from *P. apollo* by
the absence of red patches on its wings.

Family Pieridae (Whites, Sulphurs, Yellows, Brimstones)

This family is found worldwide and comprises some 1500
species, most of which are tropical. Pierids are typically
medium-sized butterflies with white, yellow or orange wings
bearing black spots. The hindwings of the males often carry
tufts of scent-bearing scales, the androconia. Many species
show strong sexual dimorphism.

The genus *Pieris* includes the Large White, *P. brassicae*
and the Small White, *P. rapae,* which are jointly referred to
as the Cabbage Whites and are annually responsible for the
depredation of numerous cabbage plots. The Small White is
the more widespread of the two and is established in North
America and Australia. Their close relative the Black-
veined White, *Aporia crataegi,* is sometimes a pest in orchards in Europe and ranges to Japan.

The beautiful Orange-tip, *Anthocharis cardamines,* which
is found from Europe to Japan, shows marked sexual
dimorphism. Only the males have the distinctive orange
patch at the apex of the forewing; in the female this is
replaced by a rather narrow and indistinct black marking.

The genus *Colias* contains the Sulphurs and Yellows and
is well represented in Europe and North America. The
commonest European species is the Clouded Yellow, *C.
croceus,* a handsome orange-yellow butterfly with black
edges to its wings. It is not resident in Britain but in some
years migrates there from North Africa and Europe. The
larva feeds on vetches.

The genus *Gonepteryx* contains the Brimstones and is
also represented in Europe and North America. These
butterflies have rather broad forewings, the apex being sub-
falcate, and have short, red antennae. The Brimstone, *G.
rhamni,* is very widely distributed in North Africa and
Europe, ranging northwards to Britain and Scandinavia.
The males are a bright sulphur yellow, while the females are
a pale greenish yellow; in both sexes there is a little orange
spot at the middle of each wing. The smooth, green caterpillar feeds on the leaves of buckthorn (*Rhamnus*). The
closely related Cleopatra, *G. cleopatra,* which is restricted
more to the warmer Mediterranean and North African
climate may be distinguished by a large, diffuse orange-red
patch on the forewing.

Subfamily Dismorphiinae

This small group of rather delicate, mostly black and white
butterflies is represented in Europe by only three species.
Perhaps the best known of these is the Wood White,
Leptidea sinapis, which inhabits open woodland. The
caterpillar feeds on various Leguminosae.

Family Danaidae (Milkweeds, Monarchs, Tigers, Crows)

These are large butterflies, generally brown in colour with
white or black markings; they can also be orange, white or
blue with brown markings, and are mostly tropical or subtropical. Most species are slow flying and have conspicuous

Right : A species of
Pararge (Satyridae).
Far right : The
Woodland Grayling,
Hipparchia fagi
(Satyridae), which
lives in wooded areas in
Europe.
Below : The Marbled
White, Melanargia
galathea (Satyridae),
which is quite a
common species in
Britain and Europe. It
gets its vernacular
name from the
distinctive wing pattern.

wing patterns and white-spotted abdomens. They are, however, distasteful to birds and other predators, since they contain poisonous substances derived from the larval food-plants, and some species can emit unpleasant odours. In both sexes the front legs are atrophied and useless for walking. Two species reach Europe, one being the Milkweed or Monarch, *Danaus plexippus*, which is an American species whose migratory habits have taken it far afield, and it is now established in the Canary Islands, Indonesia, Papua New Guinea, Australia and New Zealand. The caterpillar feeds on milkweed (*Asclepia*). Another great wanderer is the Plain Tiger, *D. chrysippus*.

Family Satyridae (Browns, Ringlets, Satyrs, Graylings)

A large worldwide group comprising about 3000 species of small to large predominantly brown butterflies. An out-standing feature of this group is the numerous little eye-spots usually present on both the upper and undersides of the wings. Many species are alpine and some have colonized the most inhospitable regions of the Arctic and Antarctic. The majority fly by day but some are most active at dusk. The caterpillars have a pair of 'tail' projections on the anal segment; they live mainly on coarse grasses, bamboo, sugar cane and palms. The large genus *Erebia*, known as Ringlets, predominates in Europe and Asia, and has representatives in North America and one in New Zealand. These include the Arctic Ringlet, *E. disa*, of Arctic Europe and North America. whose foodplant is unknown; the Mountain Ringlet, *E. epiphron*, of Britain and continental Europe, whose larva feeds on mountain grasses, and Ross's Alpine Ringlet, *E. rossi*, of North America and Asia, whose larval foodplant is also unknown.

The large genus *Pararge* is also mainly European and Asian. The Speckled Wood, *P. aegeria*, is ubiquitous in this range, and frequents shaded lanes and woodland glades. Its

bright green caterpillar, marked with yellow-edged dark green lines along the sides, feeds on various grasses. In contrast, the Meadow Brown, *Maniola jurtina*, inhabits open country up to altitudes around 1000 m (3279 ft). While the Small Heath, *Coenonympha pamphilus*, can adapt to all types of habitat from the plains to altitudes of over 2000 m (6558 ft) throughout Europe to Turkestan.

A boldly marked satyrid is the Marbled White, *Melanargia galathea*, which has black and white checkerboard wing markings; it is a common species in Europe from sea-level to over 1500 m (4920 ft).

The genus *Satyrus*, which contains the Graylings, is divided by some lepidopterists into several genera and sub-genera; its classification is thus fluid and uncertain. It contains the larger satyrids which have cryptic coloration on the underside of the wings, giving them camouflage when at rest on tree trunks or rocks. As restricted, the genus contains the Satyrs, a relatively small number of species among which is the Great Sooty Satyr, *S. ferula*, which

inhabits rocky hillsides from southern Europe to Asia Minor and the Himalayas. Two commoner species of this complex in Europe are the Woodland Grayling, *Hipparchia fagi,* and the Grayling, *H. semele*. The former frequents woodland and often rests on tree trunks, while the latter is found on heathland and open limestone country and has a characteristic habit of settling on the ground in full sun and leaning to one side to eliminate its shadow.

Family Nymphalidae (Fritillaries, Emperors, Admirals, Maps, Brush-foots)

One of the largest families which is found worldwide, comprising butterflies of considerable diversity in wing-shape and pattern. A modification found in most species of this group is the atrophied front pair of legs, which frequently have brush-like tufts. In some classifications the Danaidae and Satyridae are included in this family as sub-families. At rest, the nymphalids close their wings tightly together in a vertical position, and as these often have ragged edges and the undersides are dark-coloured, the butterflies resemble leaves. One of the best examples of leaf mimicry is found in the Indian Leaf butterfly, *Kallima inachus*, a common species in India, Pakistan, Burma and south China. When one of these butterflies settles among vegetation it virtually disappears from sight the moment it closes its wings. The outline of the wings when closed is exactly like that of a leaf, and even the leaf stalk is represented by slender appendages on the hindwings. The resemblance is further increased by lines on the wings that imitate the mid rib and lateral veins of the leaf, in addition there are small black patches on the wing that look like blemishes often found on leaves.

The numerous tawny-orange and black fast-flying Fritillaries which speed about woodland clearings and over rough ground, often at high elevations, are divided into several genera. Two of the largest European species are the Silver-washed Fritillary, *Argynnis paphia,* and the Cardinal, *Pandoriana pandora,* both of which can attain a wingspan of 8 cm (3 ins). Similar wing pattern and coloration are found in the genera *Melitaea* and *Mellicta.* The former is represented by about 40 species in Europe and Asia, but only by a single species, *M. minuta,* in North America which is found up to 3000 m (9837 ft) on mountains. The latter genus has fewer species but these are extremely variable and include numerous forms. The commonest and most widespread species is the Heath Fritillary, *M. athalia,* found in flowery meadows at low elevations from western Europe, including Britain, to Japan.

The genus *Vanessa* includes the Red Admiral, *V. atalanta,* which gets its name from the brilliant red band across its black forewing, recalling chevrons on a naval uniform. It is a migratory species ranging from Europe to Asia, and occurs in North America and Hawaii. The caterpillar feeds on nettles (*Urtica*). A related species, *V. itea,* occurs in Australia, New Zealand and the Pacific islands and is known as the Yellow Admiral or Australian Admiral. But New Zealand has its own endemic Red Admiral, *V. gonerilla,* similar to the European species in appearance except that it has four conspicuous blue spots broadly ringed with black set in the red band of the hindwing. A species sometimes placed in the genus *Vanessa* is the decorative Painted Lady, *Cynthia cardui,* which is often to be seen in Europe during the summer months feeding on buddleia flowers. It is a strong flier and a regular migrant from North Africa to Britain and northern parts of Europe. In Australia it is represented by a subspecies, *C. cardui kershawii* which differs from the European race in having the row of marginal eye spots of the hindwing blue centred.

One of the largest and most attractive nymphalids found in Europe and North America is the Camberwell Beauty, or Mourning Cloak, *Nymphalis antiopa.* Its wings are violet and brown, with a contrasting yellow border. Two other attractive European species are the Peacock, *Inachis io,* and the Small Tortoiseshell, *Aglais urticae.* Both are common and often seen in gardens, and have larvae which feed on nettles.

Members of the genus *Polygonia* can be recognized by their deeply notched borders to their wings, and the brown marbled underside of the hindwings, with a conspicuous white mark near the centre. In the Comma, *P. c-album,* this mark is in the form of a white comma or c-mark, while in the Southern Comma, *P. egea,* it is L-shaped. Both are widespread species in Europe. A close relative is the Palaearctic Map butterfly, *Araschnia levana,* which has the underside of the wings strongly patterned with thin lines and some broad bands. The genus *Charaxes* which is predominately African includes the Two-tailed Pasha, *C. jasius,* a large butterfly whose hindwings are extended into a pair of tails. It is the only charaxid found in Europe, occurring in the Mediterranean region and ranging southwards to equatorial Africa. The caterpillar feeds on the Strawberry Tree (*Arbutus*).

Above right: The Peacock butterfly, Inachis io *(Nymphalidae), which is distinguished by the four conspicuous eye-spots on its wings.*
Right: The Small Tortoiseshell, Aglais urticae *(Nymphalidae), with orange-yellow wings with black markings and blue submarginal lunules.*

Two splendid apaturids found in European woodland are the Purple Emperor, *Apatura iris*, and the Lesser Purple Emperor, *A. ilia*. These have a magnificent livery of brown with purple-blue metallic highlights, strongly iridescent in the males. *A. iris* flies high about the tops of oak trees but can be baited to ground level with a dead rabbit or other carrion. The caterpillar feeds on sallows and white poplar.

Subfamily Heliconiinae (Helicons, Passion-flower butterflies)

An exotic group comprising less than 100 species confined to the Neotropical region. Remarkably uniform in size but with a multitude of brightly coloured wing patterns, these butterflies are easily reared in captivity and are often used for genetical studies. Individual species show a wide range of variation and have recurrent forms, making identification of the species sometimes very difficult and uncertain.

Subfamily Morphinae

The remarkable brilliant blue butterflies of the genus *Morpho* are a speciality of the tropical rain-forests of South America, particularly along the Amazon. About 80 species are known, varying in wingspan from about 8 cm (3 ins) to 20 cm (8 ins). The huge wing area and the highly polished appearance of the upper surface of the wings of some of the larger species gives them a unique and spectacular appearance. The metallic gloss of the wings varies from deep blue to pale greenish white, or in some species may be brown. The butterflies are mostly fast flying and are difficult to catch. The caterpillars are usually brightly coloured and have long tufts of bristles. They feed on bamboo, leguminous, plants, and various forest trees.

Family Libytheidae (Snouts)

Only a score or so of mostly tropical species are included in this group. The butterflies are distinguished by their angular wing-shape and the enormously developed beak-like labial palpi. In this family the males have the front pairs of legs reduced as in the Nymphalidae, while the females have the normal three pairs. Only one species, the Nettle-tree butterfly, *Libythea celtis*, occurs in Europe. It is very similar in appearance to a species found in North America known as the Snout butterfly, *L. carinenta*.

Family Nemeobiidae (Metalmarks, Judies, Riodinids, Eryceinids)

A large family comprising nearly 2000 species, mostly confined to the Neotropical region. Only the Duke of Burgundy Fritillary, *Hamearis lucina* occurs in Europe, including Britain. Nemeobiids are remarkably variable in coloration and wing pattern, covering pretty well the whole of the visible colour spectrum. Many are apparently mimics of other species of butterflies and moths. The European *H. lucina* resembles a small nymphalid fritillary in general appearance, and the male also has the characteristic reduced front pair of legs. It inhabits rough, thinly wooded country, especially hillsides, and its caterpillar feeds on country flowers and most particularly on cowslips and primroses (*Primula*).

Below : The Large Copper butterfly, Lycaena dispar *(Lycaenidae), occurs from Europe to central Asia ; the British race of this species became extinct 100 years ago.*

Family Lycaenidae (Blues, Coppers, Hairstreaks)

A large family found worldwide comprising a few thousand species which are usually divided into several subfamilies, the principal ones being the Lycaeninae (Blues and Coppers) and the Theclinae (Hairstreaks). The butterflies are mostly small to medium-sized, the males usually being brightly coloured on the upperside of the wings with metallic blues or orange and coppery reds. They include the smallest butterfly in the world, the Dwarf Blue, *Brephidium barberae*, found in South Africa which has a wingspan of 14 mm (0.5 ins). The smallest butterfly found in Britain is the Small Blue, *Cupido minimus*, which has a wingspan of 19 to 25 mm (0.7 to 0.9 ins).

Lycaenid caterpillars are mostly very distinctive, being rather flattened like a woodlouse and tapered at both ends. They often have three glandular organs on the abdomen (one on the seventh segment and two on the eighth) which can be everted. These are ant-attracting honey glands and are found in the caterpillars of many species in this family which have a close relationship with ants. It has been shown that the ants seek out the secretions of the anterior gland and feed avidly on them; the posterior glands are most likely scent-glands serving as attractants. The majority of the larvae are phytophagous, feeding on the flowers, leaf buds and foliage of a wide variety of plants, but especially of legumes and lichens. A number, however, have more specialized feeding habits and are myrmecophilous, living in the nests of ants, and others are carnivorous, feeding on living scale insects, aphids and leaf-hoppers.

Some species enter into a true state of symbiosis with the ants—a mutually beneficial state in which the caterpillar provides the host ant with the secretions of its special glands, and receives protection in return. In other cases the relationship is one of parasitism rather than true symbiosis or mutualism, in that the lycaenid caterpillars feed on the pre-imaginal stages of the ants.

A familiar species throughout Europe is the Common Blue, *Polyommatus icarus*, the male of which has light violet-blue metallic-sheened wings, while the female has brown wings, only slightly flushed with blue. The caterpillar feeds on various vetches, clovers and trefoils.

The magnificent Large Copper, *Lycaena dispar*, is a widespread European species which inhabits marshes and fens. It once flourished in the East Anglian fenland of England but became extinct there just over 100 years ago, due mainly to the progressive drainage of the fens for agricultural purposes. The Scarce Copper, *Heodes virgaureae*, which prefers flowery lowland and alpine meadows, occurs from Europe to central Asia but is not found in Britain. Its caterpillar feeds on dock (*Rumex*).

The Theclinae or Hairstreaks, which get their common name from the narrow streak or row of dots on the underside of the hindwings, can generally be distinguished by the conspicuous tails that adorn their hindwings. The Green Hairstreak, *Callophrys rubi*, is one of the most common and widespread of the European species, occurring on rough scrubland and along hedgerows from sea-level up to about 2000 m (6558 ft). It is readily recognized by the nearly plain green-coloured underside of the wings, the upperside being brown. A less common European species is the Purple Hairstreak, *Quercusia quercus*, which has gleaming purple-blue coloration on the upperside of its wings. It may sometimes be glimpsed in July and August flying high about oak and ash trees, which are the larval foodplants. The slightly larger Brown Hairstreak, *Thecla betulae*, has a wingspan approaching 4 cm (1.5 ins) and a distinctive orange patch at the middle of its brown forewing, favours open woodland and occurs from central and northern Europe, including Britain, to Siberia and Korea. The caterpillar feeds chiefly on sloe and plum (*Prunus*).

Above : The Brown Hairstreak butterfly, Thecla betulae *(Lycaenidae), a woodland species found in Britain and Europe and eastwards to Asia and Korea. Like many other Hairstreaks it has the hindwings produced into short tails.*

Glossary

AEDEAGUS Part of the male genitalia; the intromittent or copulatory organ.

AESTIVATION A period of dormancy during hot or dry seasons.

AMPLEXIFORM A type of wing-coupling in which there is no frenulum and the enlarged humeral lobe of the hindwing extends beneath the forewing.

APTEROUS Without wings.

CHAETOSEMA A small, bristly sensory organ on the head of certain adult Lepidoptera, one on each side above the compound eye.

CHEMORECEPTOR A sense organ perceiving taste or smell.

CHITIN A horny substance forming an insect's skeletal structure.

CLOACA The common external opening of the reproductive and digestive systems of monotrysian Lepidoptera.

COPROPHAGOUS Feeding on dung.

CRYPTIC Protective coloration and markings facilitating concealment.

DIAPAUSE A state of dormancy, with arrested development and growth, which may occur at any stage in the life-cycle.

DIMORPHISM Occurrence of two forms in a species.

DITRYSIAN Specialized female genitalia with two genital openings: one for copulation and one for egg-laying.

DIURNAL Active in the daytime.

ECDYSIS The act of moulting.

ECOSYSTEM Ecological system formed by the interaction of living organisms and their environment.

ENDEMIC Peculiar to a particular country or region; not introduced.

EXOPORIAN A specialized monotrysian type of female genitalia peculiar to the Hepialoidea, in which there is a separate copulatory opening (as in the ditrysian type) next to the terminal genital opening.

EXUVIAE The cast skin of a larva.

FACULATIVE Optional; able to choose when to adapt to different living conditions.

FAMILY A taxonomic category comprising one genus or a number of related genera.

FRASS Caterpillar droppings or refuse, for example chewed wood.

GALEA The outer lobe of the maxilla; in Lepidoptera the two galeae are often elongated and join together to form a coiled proboscis.

GENUS A taxonomic category comprising one species or a number of closely related species.

GLABROUS Smooth and without hairs.

GNATHOS Part of the male genitalia; a pair of arm-like structures arising from the sides of the tegumen.

GYNANDROMORPH Of mixed sex; most evident in sexually dimorphic adult Lepidoptera when the left and right halves are of different sex.

HAUSTELLATE Having a proboscis adapted for sucking.

HIBERNATION A period of dormancy during winter.

HOLARCTIC REGION A faunal or zoogeographical region comprising the whole of Europe, Africa north of the Sahara, Asia to the Himalayas and America north of Mexico.

INSTAR A stage between moults in the larva.

JUGUM A lobe at the base of the forewing which extends and overlaps the base of the hindwing.

MELANISM Excessive development of black pigment.

METAMORPHOSIS The series of changes through which an insect passes in its development from the egg to the adult.

MICROPTEROUS With very small wings.

MICROPYLE The pore in the egg chorion through which the sperm enters.

MIMICRY Resemblance of one species to another living in the same habitat, or to an object such as a leaf for example.

MONOPHAGOUS Feeding on one foodplant.

MONOTRYSIAN Primitive type of female genitalia having only one genital opening, as in the suborders Monotrysia, Zeugloptera and Dacnonypha.

MYRMECOPHILOUS Living with, preying on, or mimicking ants.

NEARCTIC REGION A faunal zoogeographical region comprising North America, including Greenland; i.e. the northern part of the New World.

NEOTROPICAL REGION A faunal or zoogeographical region including the West Indies and America south of Mexico which belongs to the southern part of the New World.

NEW WORLD North and South America and the West Indies.

NOCTURNAL Active at night.

NOMENCLATURE A system of names.

OBLIGATORY Not optional.

OCELLUS A simple eye which occurs along with the compound eye in adult insects (in larvae it is usually called a stemma); also an eye-like marking, such as is found on the wings.

OLD WORLD Europe, Africa and the Indo-Australian regions.

OLIGOPHAGOUS Feeding only on a particular genus or family of foodplants.

OMMATIDIUM One of the photoreceptive units forming the compound eye.

OSTIUM The copulatory opening situated on the underside of the eighth abdominal segment of adult female Lepidoptera.

PALAEARCTIC REGION The faunal or zoogeographical region comprising Europe and most of the Old World north of the Sahara.

PARTHENOGENESIS Reproduction without fertilization by a male.

PATHOGEN A disease producing micro-organism.

PHEROMONE Chemical substance secreted by an animal to attract the opposite sex, or to influence other animals.

PHYTOPHAGOUS Feeding on plants.

POLYPHAGOUS Feeding on a variety of plants.

PROBOSCIS The spirally coiled tongue or haustellum adapted for sucking.

SACCUS Part of the male genitalia; a mid-ventral projection from the vinculum entering the body cavity.

SEXUAL DIMORPHISM Differences, especially in coloration and wingshape, between the sexes.

STEMMA A simple light sensitive ocellus found in the larva.

SYMBIOSIS A condition where different species live together in mutually beneficial partnership.

TAXONOMY The theory and practice of classifying animals and plants.

TEGUMEN Part of the male genitalia; the upper part of the ninth abdominal segment.

TUBERCLE A swelling or protuberance on the body of a larva, sometimes very bristly.

TYMPANUM An auditory membrane or ear drum.

UNCUS Part of the male genitalia; derived from the tenth abdominal segment and connected to the tegumen.

VESICA Part of the male genitalia; the terminal membranous part of the aedeagus.

WINGSPAN The measurement across the forewings from one wingtip to the other.

Bibliography

Amsel, H. G. and others *Microlepidoptera Palaearctica* (Georg Fromme, Vienna, 1965)

Bradley, J. D. and others *British Tortricoid Moths* (Ray Society at the British Museum of Natural History, London, 1973)

Brown, F. M. & Heineman, B. *Jamaica and its Butterflies* (E. W. Classey, Faringdon, England, 1972)

Common, I. F. B. & Waterhouse, D. F. *Butterflies of Australia* (Angus & Robertson, Sydney, 1972)

Corbet, A. S. & Pendlebury, H. M. *The Butterflies of the Malay Peninsular* (revised by J. N. Eliot) (E.W. Classey, Faringdon, England, 1978)

D'Abrera, B. *Moths of Australia* (Lansdowne Press, Melbourne, 1974)

D'Abrera, B. *Birdwing Butterflies of the World* (Lansdowne Press, Melbourne, 1975)

D'Abrera, B. *Butterflies of the Australian Region* (revised) (Lansdowne Press, Melbourne, 1977)

Dominick, R. B. and others *The Moths of America north of Mexico* (E. W. Classey & R. B. Publications, Faringdon, England, 1971)

Ford, R. L. E. Practical Entomology (Frederick Warne, London, 1963)

Forster, W. & Wahlfahrt, T. A. *Die Schmetterlinge Mitteleuropas* (W. Keller, Stuttgart, 1954-74)

Gaskin, D. E. *The Butterflies and common Moths of New Zealand* (Whitcombe & Tombs, Christchurch, New Zealand)

Goodden, R. *The Wonderful World of Butterflies and Moths* (Hamlyn, London, New York & Sydney, 1977)

Heath, J. and others *The Moths and Butterflies of Great Britain and Ireland* (Curwen Press, London, 1976)

Higgins, L. G. & Riley, N. D. *A Field Guide to the Butterflies of Britain and Europe* (Collins, London, 1970)

Howarth, T. G. *South's British Butterflies* (Frederick Warne, London, 1973)

Imms, A. D. and others *A General Textbook of Entomology* (Methuen, London, 1960)

Klots, A. B. *A Field Guide to the Butterflies of the U.S.A.* (Houghton Mifflin, Cambridge, Mass., 1951)

Laithwaite, E., Watson A. & Whalley P. E. S. *The Dictionary of Butterflies and Moths in colour* (Michael Joseph, London & McGraw-Hill, New York, 1975)

Lewis, H. L. *Butterflies of the World* (Harrap, London, 1974)

McCubbin, C. *Australian Butterflies* (Thomas Nelson, Melbourne, 1971)

Pinhey, E. C. G. *The Butterflies of South Africa* (Nelson, Johannesberg, 1965)

Pinhey, E. C. G. *Emperor Moths of South Africa and South Central Africa* (Struik, Cape Town, 1972)

Pinhey, E. C. G. *Moths of Southern Africa* (Tafelberg, Cape Town, 1975)

Riley, N. D. *A Field Guide to the Butterflies of the West Indies* (Collins, London, 1975)

Ross, H. H. *A Textbook of Entomology* (John Wiley, London, New York & Sydney, 1965)

Smart, P. *The Illustrated Encyclopedia of the Butterfly World* (Hamlyn, London, New York & Sydney, 1975)

South, R. *The Moths of the British Isles* (Frederick Warne, London, 1961)

Spuler, A. *Die Schmetterlinge Europas* (E. Schweizerbartsche, Stuttgart, 1908–10)

Verity, R. *Le Farfalle diurne d'Italia* (Marzocco, Florence, 1940–53)

Wigglesworth, V. B. *The Life of Insects* (Weidenfeld & Nicolson, London, 1964)

Williams, *A Field Guide to the Butterflies of Africa* (Collins, London, Toronto, Sydney, Aukland, 1969)

Index